Welding Robots

J. Norberto Pires, Altino Loureiro and
Gunnar Bölmsjo

Welding Robots

Technology, System Issues and Applications

With 88 Figures

Springer

J. Norberto Pires, PhD
Altino Loureiro, PhD
Mechanical Engineering Department
Robotics Laboratory
University of Coimbra
Polo II Campus
3030 Coimbra
Portugal

Gunnar Bölmsjo, PhD
Mechanical Engineering Department
Lund Institute of Technology
Sweden

British Library Cataloguing in Publication Data
Pires, J. Norberto
Welding robots : technology, systems issues and applications
1. Welding - Automation 2. Robots
I. Title II. Loureiro, Altino III. Bolmsjo, Gunnar
671.5'2

ISBN-13: 978-1-84996-973-4 e-ISBN 1-84628-191-1 Printed on acid-free paper
e-ISBN-13: 978-1-84628-191-4

Softcover reprint of the hardcover 1st edition 2006

9 8 7 6 5 4 3 2 1

Springer Science+Business Media
springeronline.com

Dedicated to the memory of my father Joaquim
and to Dina, Rita, Beatriz and Olímpia.
J. Norberto Pires

Dedicated to my wife Marília and my sons André and Joana
Altino Loureiro

Dedicated with love to Ziona,
Rebecka, Natalie and Daniela.
Gunnar Bolmsjö

Foreword

Industrial robots are essential components of today's factory and even more of the factory of the future. The demand for the use of robots stems from the potential for flexible, intelligent machines that can perform tasks in a repetitive manner at acceptable cost and quality levels. The most active industry in the application of robots is the automobile industry and there is great interest in applying robots to weld and assembly operations, and material handling.

For the sake of competitiveness in modern industries, manual welding must be limited to shorter periods of time because of the required setup time, operator discomfort, safety considerations and cost. Thus, robotic welding is critical to welding automation in many industries. It is estimated as much as 25% of all industrial robots are being used for welding tasks.

Robotic welding is being initiated to satisfy a perceived need for high-quality welds in shorter cycle times. The first generation of robotic welding system was a two-pass weld system, where the first pass is dedicated to learning the seam geometry followed by the actual tracking and welding in the second pass. The second generation of welding systems, on the other hand, track the seam in real-time, performing simultaneously the learning and the seam tracking phases. The third generation of welding systems not only operates in real-time but also learns the rapid changing in seam geometries while operating within unstructured environments. Flexibility was achieved with this third generation of welding systems but at the expenses of a considerable amount of programming work of high skilled people in system's integration directed to specific applications. However, availability and agility are additional key issues in modern manufacturing industries, demanding new welding systems incorporating these features as well, revealing in this way the flexibility of the system to the normal operator without the need of extra skills from him.

This book covers up-to-date and relevant work in the area of third generation of robotic welding systems with availability and agility features. The principal welding processes are reviewed from the point of view of their automation. A distributed system's approach is followed for the integration of the different components and software of the welding cell and its integration within the global production system. Particular emphasis is given to the availability and agility to the end user. Application examples demonstrating step-by-step the system's integration design clarify the relevant aspects to the interested reader.

The authors have made a strong-minded effort to set their work in the context of international robotic arc welding research. The mix of specific research issues and the review of broader research approaches make this a particularly welcome contribution.

This book is directed towards readers who are interested in developing robotic welding applications, and in particular to perform system integration. Although this work is presented in the context of arc welding, the issues related to system integration are general in nature and apply to other robotic applications as well. This book constitutes a valuable source of the kind of information on robotic welding that result of years of experience, making it suitable as well for the decision maker, the application engineer, the researcher, the technician, and the student.

José Sá da Costa
Mechanical Engineering Department
Superior Technical Institute (IST)
Technical University of Lisbon
Portugal

Preface

Modern manufacturing faces two main challenges: more quality at lower prices and the need to improve productivity. Those are the requirements to keep manufacturing plants in developed countries, facing competition from the low-salary regions of the world. Other very important characteristics of the manufacturing systems are flexibility and agility of the manufacturing process, since companies need to respond to a very dynamic market with products exhibiting very short life-cycles due to fashion tendencies and worldwide competition. Consequently, manufacturing companies need to respond to market requirements efficiently, keeping their products competitive. This requires a very efficient and controlled manufacturing process, where focus is on automation, computers and software. The final objective is to achieve semi-autonomous systems, *i.e.*, highly automated systems that work requiring only minor operator intervention.

Robotic welding is one of the most successful applications of industrial robot manipulators. In fact, a huge number of products require welding operations in their assembly processes. Despite all the interest, industrial robotic welding evolved only slightly and is far from being a solved technological process, at least in a general way. The welding process is complex, difficult to parameterize and to monitor and control effectively. In fact, most of the welding techniques are not fully understood, namely the effects on the welding joints, and are used based on empirical models obtained by experience under specific conditions. The effects of the welding process on the welded surfaces are currently not fully known. Welding can in most cases impose extremely high temperatures concentrated in small zones. Physically, that makes the material experience extremely high and localized thermal expansion and contraction cycles, which introduce changes in the materials that may affect its mechanical behavior along with plastic deformation. Those changes must be well understood in order to minimize the effects.

The majority of industrial welding applications benefit from the introduction of robot manipulators, since most of the deficiencies attributed to the human factor is removed with advantages when robots are introduced. This should lead to cheaper products since productivity and quality can be increased, and production costs and manpower can be decreased. Nevertheless, when a robot is added to a welding setup the problems increase in number and in complexity. Robots are still difficult to use and program by regular operators, have limited remote facilities and programming environments, and are controlled using closed systems and limited software interfaces.

The present book gives a detailed overview of Robotic Welding at the beginning of the twenty-first century. The evolution of robotic welding is presented, showing to the reader what were the biggest steps and developments observed in the last few years. This is presented with the objective of establishing the current state-of-the-art in terms of technologies, welding systems, software and sensors. The remaining issues, *i.e.*, the issues that remain open are stated clearly, as a way to motivate the readers to follow the rest of the book which will make contributions to clarify most of them and help to solve a few.

To do that, a good chapter on "Welding Technology" is presented, describing the most important welding techniques and their potential and requirements for automation using robot manipulators. This chapter includes recent results on robotic welding processes, which can constitute a good source of information and practical examples for readers.

A good revision with current research results on "Sensors for Welding Robots" used on robotic welding is also presented. This includes sensors for seam tracking, quality control and supervision. This chapter includes all system requirements necessary to use those sensors and sensing techniques with actual robot control systems. Hardware and software interfaces are also covered in detail.

A good revision on available welding systems, including hardware and software, clarifying their advantages, and drawbacks is also necessary to give to the reader a clear picture of the area. The book includes a chapter on "Welding Robots: System Issues", which covers recent state-of-the-art of industrial robotic welding systems currently available in industry and university laboratories.

Finally, a few industrial applications using the presented techniques and systems is presented. The present book includes a chapter on "Robotic Welding: Application Examples", where a few selected applications are described in detail including aspects related to software, hardware, system integration and industrial exploitation. This chapter uses actual robots, but it is presented in a general way so that the interested reader can easily explore his interests.

Conclusions stating what was presented and what are the next challenges, guiding the reader to what are the next required developments, is presented at the end of the book. A good collection of references is also presented, to enable the reader to explore further from the literature.

J. Norberto Pires, Coimbra, 2005

Contents

List of Figures

1

Introduction and Overview

1.1 Introduction

Actual market conditions are only compatible with small/medium batch manufacturing, due to strong competition and dynamical behavior of the market. In those conditions, robotic production setups exhibit the best "cost per unit" performance if compared with manual work and with hard automated setups (Figure 1.1) [1]. Consequently, near future requires powerful and more flexible machines in order to handle requests from small businesses, which need more remote interfaces, powerful programming languages, force control, powerful Advanced Programming Interfaces (APIs) for high level programming, *etc.* That means exposing to the user the flexibility stored inside the manufacturing robotic machines, as a result of several decades of engineering, which is currently barely used.

What makes robotics so interesting is that it is a science of ingenious devices, constructed with precision, powered by a permanent power source, and flexible from the programming point of view. That does not mean necessarily open source, but instead the availability of powerful APIs, and *de facto* standards both for hardware and software, enabling access to system potentialities without limitations. This is particularly necessary on research environments, where a good access to resources is needed in a way to implement and test new ideas. If that is available, then a system integrator (or even a researcher) will not require open source software, at least for the traditional fields of robotics (industrial robot manipulators and mobile robots). In fact, that could also be very difficult to achieve since those fields of robotics have decades of engineering efforts, achieving very good results and reliable machines, which are not easy to match. That open source issue is nevertheless very important for the emerging robotics research (like humanoid robotics, space robotics, robots for medical use, *etc.*) as a way to spread and accelerate development (Figure 1.2).

Industrial Robotic Welding is by far the most popular application of robotics worldwide [6]. In fact, there is a huge number of products that require welding operations in their assembly processes. The car industry is probably the most important example, with the spot and MIG/MAG welding operations in the car body workshops of the assembly lines. Nevertheless, there are an increasing number of smaller businesses, client oriented, manufacturing small series or unique products designed for each client. These users require a good and highly automated welding process in a way to respond to client needs in time and with high quality. It is for these companies that the concepts of *Agile Production* [7],[8] apply the most, obviously supported by flexible manufacturing setups. Despite all this interest, industrial robotic welding evolved slightly and is far from being a solved technological process, at least in a general way. The welding process is complex, difficult to parameterize and to effectively monitor and control [1]-[7]. In fact, most of the welding techniques are not fully understood, namely the effects on the welding joints, and are used based on empirical models obtained by experience under specific conditions. The effects of the welding process on the welded surfaces are currently not fully known. Welding can in most cases (*i.e.* MIG/MAG welding) impose extremely high temperatures concentrated in small zones. Physically, that makes the material experience extremely high and localized thermal expansion and contraction cycles, which introduce changes in the materials that may affect its mechanical behavior along with plastic deformation [9]-[11]. Those changes must be well known in order to minimize the effects.

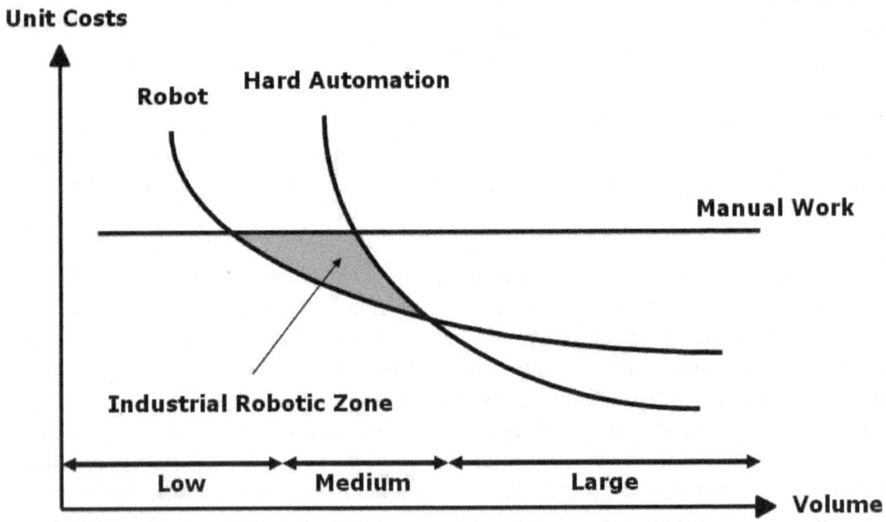

Figure 1.1. Industrial robot zone

Using robots with welding tasks is not straightforward and has been a subject of various R&D efforts [12]-[16]. And that is so because the modern world produces a huge variety of products that use welding to assemble some of their parts. If the percentage of welding connections incorporated in the product is big enough, then some kind of automation should be used to perform the welding task. This should lead to cheaper products since productivity and quality can be increased, and production costs and manpower can be decreased [17]. Nevertheless, when a robot is added to a welding setup the problems increase in number and in complexity. Robots are still difficult to use and program by regular operators, have limited remote facilities and programming environments, and are controlled using closed systems and limited software interfaces [18]-[22].

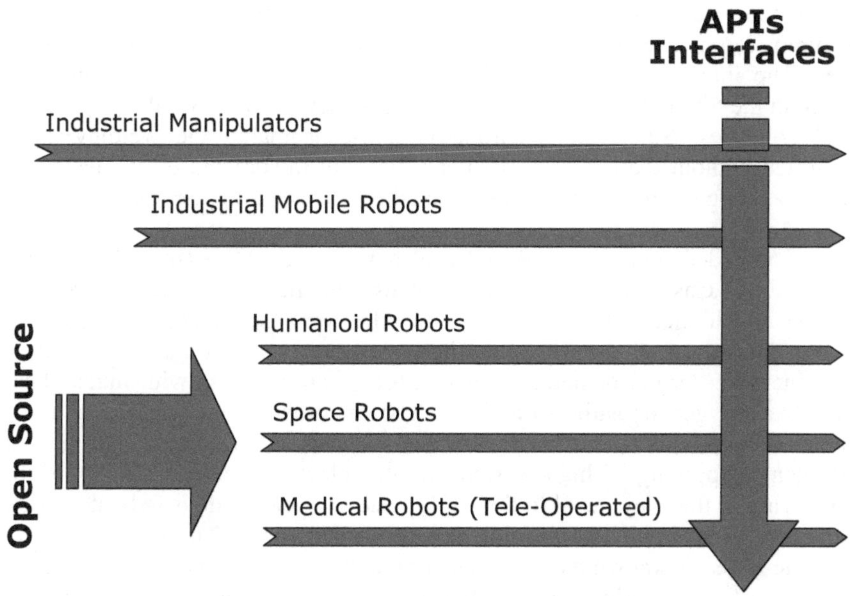

Figure 1.2. Traditional and modern fields in robotics research: where is open source needed?

In this book, most of these problems are addressed in detail along with a comprehensive presentation and discussion of a laboratory system built with the main objective of being a test bed for welding experiments. Our experience with the system shows that it has potentialities for industrial utilization, and in fact that idea is explored in the book, using industrial partner test-cases. For that purpose mainly industrial equipment was selected in designing the system, as a way to facilitate its industrial exploitation. The book also addresses aspects of system programming and welding parameterization, which constitute one of the main contributions of the book.

1.1.1 Why Robotic Welding and a CAD Programming Interface?

Automation of the welding process is a very challenging area of research in the fields of robotics, sensor technology, control systems and artificial intelligence. This book discusses the automation of the welding process taking as an example the arc welding process. Although there's a huge number of welding processes, usually suited for a particular type of application, arc welding is used in nearly all applications in the metal manufacturing industry. The two most common types of arc welding processes are the gas shielded tungsten arc welding (GTAW) and the gas shielded metal arc welding (GMAW) processes.

The gas shielded tungsten arc welding process (GTAW), also known as tungsten inert gas (TIG), is a welding process where the arc is created between a non-consumable electrode and the work metal. The process is shielded from contamination by the atmosphere using an inert gas, usually argon or a mixture of gases. The intense heat, generated by the electric arc produced by an electric current in the 50 to 700 A range, melts the work metal and allows the joining as the metal solidifies. Since the electrode is non-consumable the welding can be performed without the addition of filler metal, but in some cases a filler metal is used depending on the requirements established for the particular join.

The gas shielded metal arc (GMAW), also known as MIG (Inert Gas Metal) / MAG (Active Gas Metal) welding process, uses the heat of the electric arc to melt the consumable electrode wire and the metallic components to be welded. Figure 1.3 illustrates the welding principle. The fusion is carried out under the protection of an inert gas (argon or helium), or mixture of an inert gas with much cheaper gases like oxygen or carbon dioxide (CO_2), in order to prevent the pernicious contamination with some gases of the atmosphere (oxygen, nitrogen and hydrogen). Applying a high current to the electrode causes its tip to melt transferring in this way metal to the work-piece. The electrode is fed automatically to the arc using a coil that unfolds at a controlled speed. The rate at which the electrode is fed is known as *wire feed rate*, and is one parameter of fundamental importance for controlling this welding process. Depending on the magnitude of the electrode current and voltage, along with the type of gas and size of the electrode, four different types of metal transfer modes can be obtained: spray, short-circuiting, globular and pulsed transfer.

A complete description of these and other current welding processes will be presented in Chapter 2. Nevertheless, the brief description above makes it easy to conclude that a good quality weld relies on the welder's experience and skill. The experienced and skilled manual welder is able to select the welding process parameters based on similar cases previously encountered. In particular, he is able to:

1. Select the type of shielding gas, the type and diameter of wire to use, and the initial current and voltage settings more suitable for the case in hand.
2. Adjust continuously the process variables by looking to the molten pool or by listening to the sound produced by the arc.
3. Maintain the torch in the correct position with precision and stability, which is fundamental for a good and constant weld.

Consequently, the task of automating the welding operation is to reproduce the experienced and skilled manual welder in terms of positioning the welding torch, and controlling the welding parameters. That means availability of databases that register known cases, from where initial conditions can be selected, along with type of shielding gases and wires. That means also the capacity to observe the ongoing process and adjust or adapt the controlling parameters in accordance with the desired results. And finally, the possibility of holding the welding torch and move it in a precise and controlled way. Therefore, as previously mentioned, automating the welding process is a mixture of robotics research, control systems research, sensor research, sensor fusion and artificial intelligence.

Let's consider for example the MIG/MAG welding process. The stability of the welding process is very sensitive to the main welding parameters, especially *current*, *voltage*, *welding speed*, *stick-out* (length of wire out of the contact tube), *shielding gas* and *arc length* [24]. A small change in the distance between the welding torch and the component being welded may produce a considerable variation in the current and in the voltage. Current, voltage and shielding gas influence the transfer mode of melted filler metal to the component being welded, affecting the quality of the welds [25]. If the electric arc is unstable, defects like bad penetration profile, undercut or excessive spatter may occur.

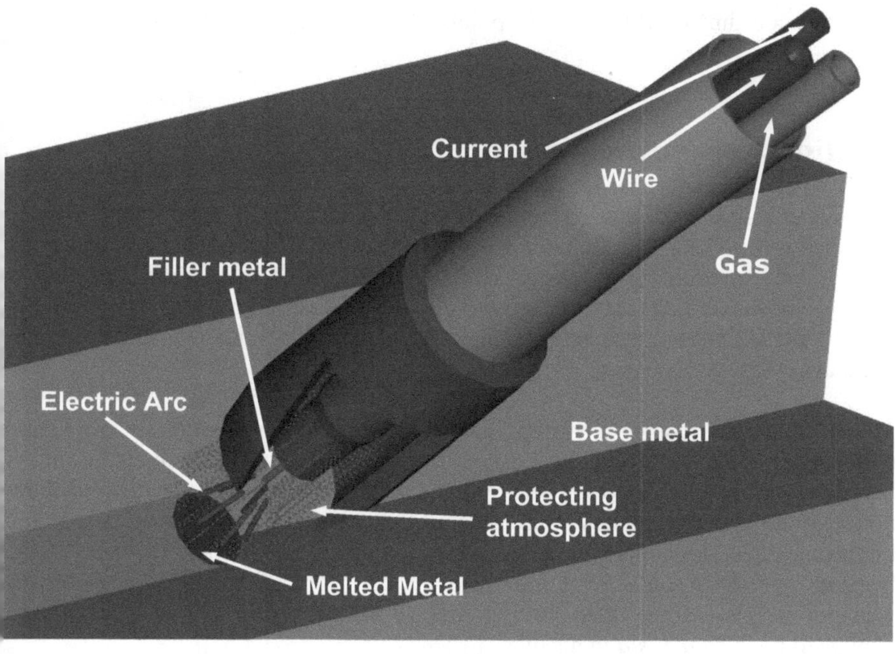

Figure 1.3. MIG/MAG welding principle

As the weld bead shape may be closely related with the welding parameters, databases for MIG/MAG welding process have been developed, such as that of *The*

Welding Institute – UK [26]. In these databases the input data is generally the type of weld (butt weld or fillet weld), the welding position (flat, horizontal, vertical or overhead), wire diameter and the plate thickness or eventually the leg length in the case of fillet welds. The output data is usually the welding parameters (namely, *current*, *voltage*, *welding speed* and *number of weld beads/layers*). Using databases of this type with a computer, the selection of the welding parameters may be performed automatically. Even the selection of the wire diameter may be carried out automatically as a function of the thickness of the components, or stay for free user selection being an input parameter.

It might be expected that with this information in the computer, having a CAD model of the component to be welded, the system would be able to select the welding data for each weld and send these data to the robotic welding system. Though it seems easy to achieve this goal in the case of single welds, some data are missing in the available databases for the case of welds with multiple layers. In fact, in this case the position of the torch in each layer needs to be indicated to the robot.

Since for the majority of the companies that produce multi-layer welds there is only a small number of distinct welds, then it is not hard to fill up the database for their particular case. Consequently, using this method it is easy to carry out the off-line programming of the components to be welded, it being only necessary to adjust the coordinates of the process points in the first specimen to be welded.

1.2 Historical Perspective

Welding is a skill used to manufacture, produce, construct and repair metal objects. In fact this skill can also be used to join other type of materials, but this book focuses only on welding processes used to join metal objects, where this skill is critical for several areas of activity like defense, aerospace, shipbuilding, transportation, building and bridge construction, industrial apparatus and consumer products.

The word "*robot*" comes from the Czech "*robota*" that means tireless work, and was used for the first time in 1921 by the novelist *Karel Capek* in his novel "*Rossum's Universal Robots*". But robotics was in the head of the most brilliant minds of our common history, since most of them took time to imagine, design and build machines that could mimic some of the human capabilities. It is one of the biggest dreams of man, to build obedient and tireless machines, capable of substituting man doing their boring and repetitive work. An idea very well explained by *Nicola Tesla* in his diary [4]:

> *"... I conceived the idea of constructing an automaton which would mechanically represent me, and which would respond, as I do myself, but, of course, in a much more primitive manner, to external influences. Such an*

> *automaton evidently had to have motive power, organs for locomotion, directive organs, and one or more sensitive organs so adapted as to be excited by external stimuli ...".*

In the next two sub-sections a brief overview of the history of both welding and robotics will be given.

1.2.1 Welding

Welding is also an ancient craft that combines art, science and human skill. It can be traced back to around 3000 BC, with the *Sumerians* and the *Egyptians*. The *Sumerians* used to made swords with parts joined by hard soldering. The *Egyptians* found that after heating iron, it was much easier to work with, or apply "*pressure*" welding or "*solid-state*" welding just by hammering the parts to join. These are the first recorded welding procedures. Several objects were found in tombs, excavations, *etc.*, indicating the use of several welding techniques, like "*pressure*" (hammering) welding, applied with several metal materials (gold, iron, bronze, copper, *etc.*), in those ancient times.

In the sixteenth century these basic welding techniques were well known but not used to any great extent. In 1540, the Italian Engineer *Vannoccio Biringuccio* explains in his book *"The Pirotechnia"*, published in Venice [35], that welding "*seems to me an ingenious thing, little used, but of great usefulness*", and he continues:

> *"the secret of welding a fracture of a saw, a sickle, or a sword, resides in taking some low silver, borax or crushed glass and embracing the fracture with a pair of hot tongs and closing so tight till the welding leans out and so cools"*

During these middle ages, the art of blacksmithing was further developed and it was possible to produce many items of iron welded by hammering. It was not until the nineteenth century that welding, as we know it today, was invented.

In the nineteenth century and early twentieth century several discovers in the field of electricity and magnetism, but also in metallurgy, heat transfer and thermodynamics, anticipated the amazing evolution done on welding during the twentieth century. In 1800 *Alessandro Volta* finds a way to store energy in his "*voltaic cell*" (battery), just by connecting two dissimilar metals using a moistened substance. This was the first step to use electricity effectively. One year later, in 1801, the eminent English scientist *Sir Humphrey Davy*, demonstrated how to generate an electric arc between two carbon electrodes. The same scientist discovered magnesium and proved the existence of aluminum (finally discovered in 1827 by *Friederich Wöler*), both in 1808. He also discovered acetylene in 1836. In the mid-nineteenth century, the electric generator was invented and arc lighting became popular. During the late 1800s, gas welding and cutting was developed.

Arc welding with the carbon arc and metal arc was developed and resistance welding became a practical joining process.

In 1881, *Auguste De Meritens*, working in the Cabot Laboratory (France), used the heat of an electric arc for joining lead plates for storage batteries. The process was patented in France by his Russian protégé, *Nikolai N. Benardos*, which also secured, with a Russian colleague named *Stanislaus Olszewski*, a British patent in 1885 and an American patent in 1887. The patents show an early electrode holder. This was the beginning of carbon-arc welding. *Bernardos'* efforts were restricted to carbon arc welding, very popular in the following 20 years, although he was able to weld iron as well as lead.

In 1890, C.L. Coffin registered the first U.S. patent for an arc welding process using a metal electrode. This was the first record of a welding process where the metal, melted from the electrode, was carried across the arc to deposit filler metal in the joint to make a weld. This neat idea of transferring metal across an arc was presented, about the same time, by the Russian *N.G. Slavianoff*, to cast metal in a mold. Interesting coincidence.

Around 1900, *Strohmenger* introduced a coated metal electrode in England. The coating, made of clay or lime, was very thin but sufficient to provide a more stable arc. *Oscar Kjellberg* and the *ESAB Company*, both from Sweden, invented a covered or coated electrode during the period 1907 to 1914. Stick electrodes were produced by dipping short lengths of bare iron wire in thick mixtures of carbonates and silicates, and allowing the coating to dry.

Meanwhile, resistance welding processes were also developed, including spot welding, seam welding, projection welding and flash butt welding. *Elihu Thompson* originated resistance welding in the nineteenth century: his patents are dated from 1885 to 1900. In 1903, a German named *Goldschmidt* invented thermite welding that was first used to weld railroad rails. The first automobile body spot welded was built by *E.G. Budd* in Phyladelphia (USA) in 1912.

Gas welding and cutting were perfected during this period as well. The production of oxygen and later the liquefying of air, along with the introduction of a blow pipe, or torch, in 1887, helped the development of both welding and cutting. Before 1900, hydrogen and coal gas were used with oxygen. However, in about 1900 a torch suitable for use with low-pressure acetylene was developed.

World War I brought a tremendous demand for armament production, which means huge production of heavy and very dissimilar metal parts. Consequently, welding was pressed into service as a way to respond to those production demands, giving the opportunity to several companies to appear, both in America and Europe, and manufacture the necessary welding machines and electrodes.

Immediately after the war in 1919, 20 members of the *Wartime Welding Committee* of the *Emergency Fleet Corporation* under the leadership of *Comfort*

Avery Adams, founded *the American Welding Society*, a nonprofit organization dedicated to the advancement of welding and allied processes.

Alternating current, invented in 1882 by *Nicola Tesla*, was applied to welding for the first time by *C.J. Holslag* in 1919. However it did not became popular, for welding, until the 1930s when the heavy-coated electrodes become generally used.

In 1920, automatic welding was invented by *P.O. Nobel* of the *General Electric Company*. It was used to build up worn motor shafts, worn crane wheels, and rear axle housings for the automobile industry. This process utilized bare electrode wire operated on direct current and utilized arc voltage as the basis of regulating the feed rate.

During the 1920s, various types of welding electrodes were developed, with a considerable controversy about the advantage of the heavy-coated rods *vs* light-coated rods. By 1930, covered electrodes were widely used. Welding codes appeared which required higher-quality weld metal, which increased the use of covered electrodes.

Also during the 1920s there was considerable research in trying to shield the arc and weld area by externally applied gases. The atmosphere of oxygen and nitrogen in contact with the molten weld metal caused brittle and sometime porous welds. Research work was done utilizing gas shielding techniques. *Alexander and Langmuir* did some exploratory work in chambers using hydrogen as a welding atmosphere. They first utilized two electrodes of carbon, but changed later to tungsten. The hydrogen was also changed to atomic hydrogen near the arc, because the flame produced was more intense than the molecular form produced flame, and as intense as an oxyacetylene flame. This then became known as the atomic hydrogen welding process. Atomic hydrogen never became popular but was used during the 1930s and 1940s for special applications of welding and later on for welding of tool steels.

H.M. Hobart and *P.K. Devers* were doing similar work but using atmospheres of argon and helium. Their patents (1926) were the predecessors of the gas tungsten arc welding process, because they showed how to carry out arc welding utilizing gas supplied around the arc. They also showed welding with a concentric nozzle and with the electrode being fed as a wire through the nozzle. This was the predecessor of the gas metal arc welding process (GMAW), which was developed only 20 years later.

Stud welding was developed in 1930 at the New York Navy Yard, specifically for attaching wood decking over a metal surface. Stud welding became popular in the shipbuilding and construction industries.

The automatic process that became popular was the submerged arc welding process. This "*under powder*" or smothered arc welding process was developed by the *National Tube Company* for a pipe mill at McKeesport, Pennsylvania. It was designed to make longitudinal seams in pipe. The process was patented by *Robinoff* in 1930 and was later sold to *Linde Air Products Company*, where it was renamed *Unionmelt®* welding. Submerged arc welding was actively used during the 1938 defense buildup in shipyards and in ordnance factories. It is one of the most productive welding processes and remains popular today.

Gas tungsten arc welding (GTAW) had its beginnings in an idea by *C.L. Coffin* to weld in a non-oxidizing gas atmosphere, which he patented in 1890. The concept was further refined in the late 1920s by *H.M. Hobart*, who used helium for shielding, and *P.K. Devers*, who used argon. This process was ideal for welding magnesium and also for welding stainless steel and aluminum. It was perfected in 1941, patented by *Meredith*, and named *Heliarc®* welding. It was later licensed to *Linde Air Products*, where the water-cooled torch was developed. The gas tungsten arc welding process has become one of the most important gas arc welding processes.

The gas shielded metal arc welding (GMAW) process was successfully developed at the *Battelle Memorial Institute* in 1948 under the sponsorship of the *Air Reduction Company*. This development utilized the gas shielded arc, similar to the gas tungsten arc, but replaced the tungsten electrode with a continuously fed electrode wire. One of the basic changes that made the process more usable was the small-diameter electrode wires and the constant-voltage poser source (a principle patented earlier by *H.E. Kennedy*). The initial introduction of GMAW was for welding nonferrous metals. The high deposition rate led users to try the process on steel, but since the cost of inert gas was relatively high at the time, the cost savings were not immediately evident.

In 1953, *Lyubavskii* and *Novoshilov* announced the use of welding with consumable electrodes in an atmosphere of CO_2 gas. The CO_2 welding process immediately gained favor since it utilized equipment developed for inert gas metal arc welding, but could now be used to perform more economical welds with steels. Since the CO_2 arc is a hot arc requiring fairly high currents for larger electrodes, the process only became widely used with the introduction of smaller-diameter electrode wires and more efficient power supplies. Those power supplies used the short-circuit arc variation, also known as *Micro-wire®*, *short-arc*, or *dip transfer welding*, all of which appeared late in 1958 and early in 1959. This variation allowed welding on thin materials and every position, and soon became the most popular of the gas metal arc welding process variations.

Another variation was the use of inert gas with small amounts of oxygen that provided the spray-type arc transfer. It became popular in the early 1960s.

A more recent variation is the use of pulsed current. The current is switched from a high to a low value at a rate of once or twice the line frequency (50 Hz in Europe).

Soon after the introduction of CO_2 welding, a variation utilizing a special electrode wire was developed. This wire, described as an inside-outside electrode, was tubular in cross section with the fluxing agents on the inside. The process was called *Dualshield®*, which indicated that external shielding gas was utilized, as well as the gas produced by the flux in the core of the wire, for arc shielding. This process, invented by *Bernard*, was announced in 1954, but was patented in 1957, when the *National Cylinder Gas Company* reintroduced it.

In 1959, an inside-outside electrode was produced which did not require external gas shielding. The absence of shielding gas gave the process popularity for non-critical work. This process was named *Innershield®*.

The *electroslag* welding process was announced by the Soviets at the Brussels World Fair in Belgium in 1958. It had been used in the Soviet Union since 1951, but was based on work done in the United States by *R.K. Hopkins*, who was granted patents in 1940. The *Hopkins* process was never used to a very great degree for joining. The process was perfected and equipment was developed at the *Paton Institute Laboratory* in Kiev, Ukraine, and also at the *Welding Research Laboratory* in Bratislava, Czechoslovakia. The first production use in the U.S. was at the Electromotive Division of the General Motors Corporation in Chicago, where it was called the Electro-molding process. It was announced in December 1959 for the fabrication of welded diesel engine blocks. The process, and its variation using a consumable guide tube, is used for welding thicker materials.

The *Arcos Corporation* introduced another vertical welding method, called *Electrogas*, in 1961. It utilized equipment developed for *electroslag* welding, but employed a flux-cored electrode wire and an externally supplied gas shield. It is an open arc process since a slag bath is not involved. A newer development uses self-shielding electrode wires and a variation uses solid wire but with gas shielding. These methods allow the welding of thinner materials than can be welded with the *electroslag* process.

Robert F. Gage invented plasma arc welding in 1957. This process uses a constricted arc or an arc through an orifice, which creates an arc plasma that has a higher temperature than the tungsten arc. It is also used for metal spraying and for cutting.

The electron beam welding process, which uses a focused beam of electrons as a heat source in a vacuum chamber, was developed in France. *J.A. Stohr* of the French Atomic Energy Commission made the first public disclosure of the process on November 23, 1957. In the United States, the automotive and aircraft engine industries are the major users of electron beam welding.

Friction welding, which uses rotational speed and upset pressure to provide friction heat, was developed in the Soviet Union. It is a specialized process and has applications only where a sufficient volume of similar parts is to be welded because of the initial expense for equipment and tooling. This process is called inertia welding.

The company *TWI* (Cambridge, England) developed in 1991 the new and impressive *Friction Stir Welding Process* in its laboratory. This process is considerably different from the rotary technology whereby a hard, non consumable, cylindrical tool causes friction, plasticizing two metals into a *solid-state bond*. This process does not require any shielding gas or filler metal, produces good quality welds for at least aluminum series 2XXX, 6XXX and 7XXX, and was used successfully to weld the impressive fuel tank of the *Space Shuttle* (NASA).

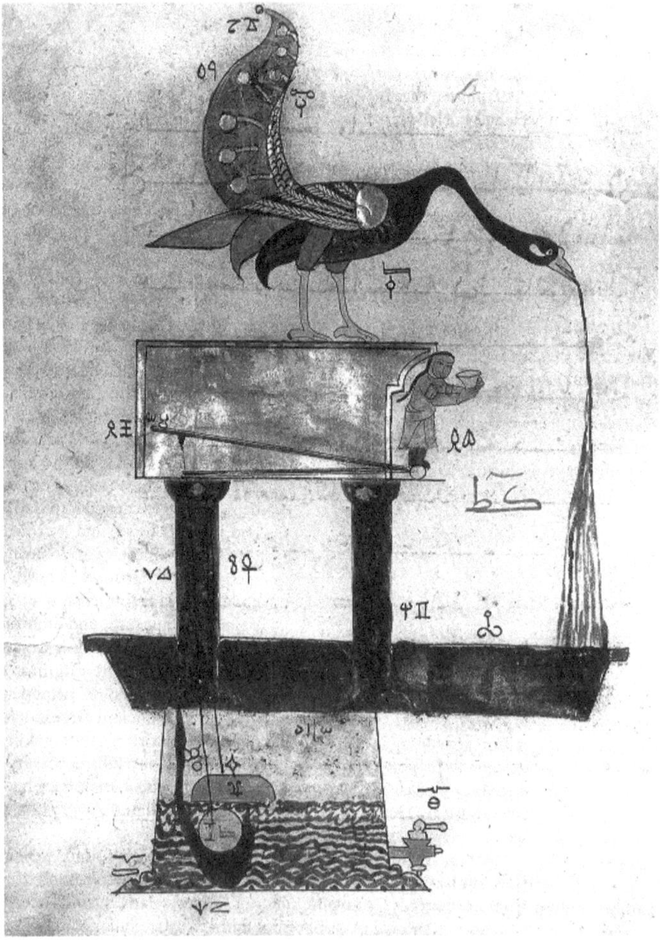

Figure 1.4. A Greek design adapted by *al-Jazari* for a garden animated hand-washer

Laser welding is one of the newest processes. The laser was originally developed at the Bell Telephone Laboratories as a communications device. Because of the tremendous concentration of energy in a small space, it proved to be a powerful heat source. It has been used for cutting metals and nonmetals. Continuous pulse equipment is available. The laser is finding welding applications in automotive metalworking operations. The first automotive production application of laser welds was conducted by General Motors, using two 1.25 KW CO_2 lasers for welding valve assemblies used in the emission control systems.

1.2.2 Robotics

If we remember the long history of robotics there are a few things we can learn in order to understand our present situation. Robotics can be traced back to 270 BC, in ancient *Greece*, to the water clocks of the Civil Engineer *Ctecibius*. His work had followers like *Phylo of Byzantium*, author of the book "*Mechanical Collection*" (200 BC), and also *Hero of Alexandria* (85 BC) *and Marcus Vitruvius* (25 BC). In the twelfth century, the Arabian *Badías-zaman al-Jazari* (1150-1220) recollected some of the Greek developments in the book "*The Science of the Ingenious Devices*" [1] (Figure 1.4), and that is how they reached our time. In those early times the problem was about mechanics, about how to generate and transmit motion. So it was mainly about mechanisms, ingenious mechanical devices [1],[2].

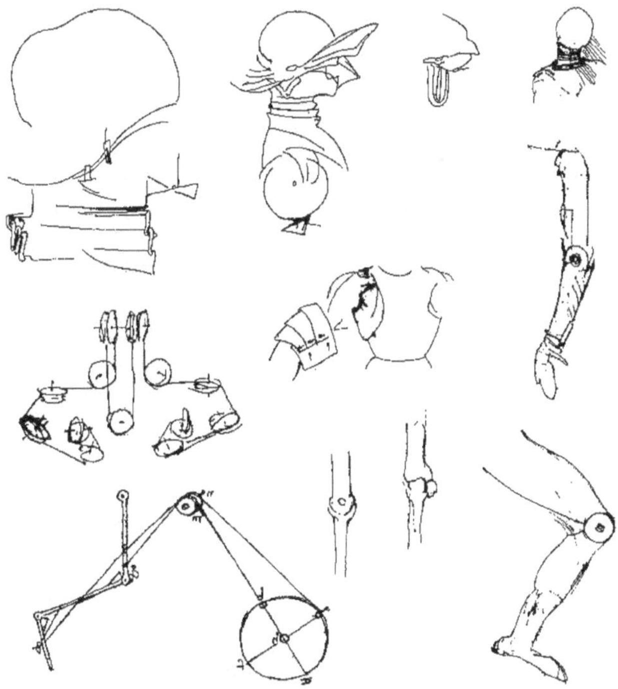

Figure 1.5. Leonardo's studies for a humanoid robot

Then in the fifteenth century, *Leonardo da Vinci* showed indirectly that the problem at the time was mainly the lack of precision and of a permanent power source. He designed a lot of mechanisms to generate and transmit motion, and even some ways to store small amounts of mechanical energy [3]. But he didn't have the means to build those mechanisms with enough precision and there was no permanent power source available (pneumatic, hydraulic or electric). Maybe that was why he didn't finish his robot project [1],[2], a fifteenth century knight robot (Figure 1.5) intended to be placed in the "*Salle delle Asse*" of the *Sforza* family castle (Milan, Italy). It wasn't good enough. Or it was a so revolutionary idea for the time that he thought that maybe it was better to make it disappear [1],[2].

And then there was the contribution of *Nicola Tesla* at the turn of the nineteenth century. He thought of using *Henrich Hertz's* discovery of radio waves (following the work of *James Clerk Maxwell* about electromagnetic phenomena) to command an automata. He built one (Figure 1.6) to demonstrate his ideas and presented it in the *Madison Square Garden* (New York, USA) in 1905 [1],[4]. The problem there was that machine intelligence was missing. Robots should be able to do pre-programmed operations, and show some degree of autonomy in order to perform the desire tasks. When that became available, robots developed rapidly and the first industrial one appeared in the beginning of the 1970s and became a multi-million dollars business.

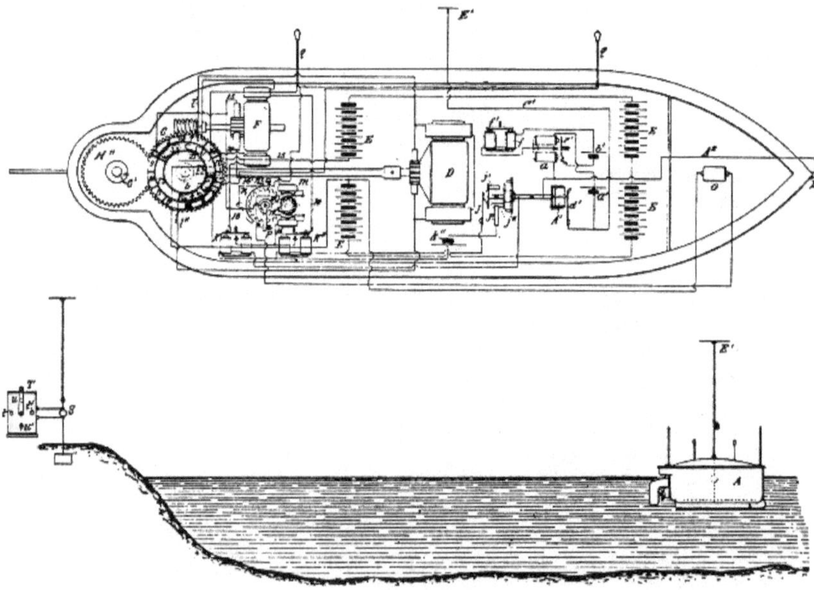

Figure 1.6. Nicola Tesla's remote controlled miniature submarine

Since then, evolution was not as fantastic as it could have been, since there was a lot to do and the available machines were sufficiently powerful to handle the requested jobs. Manufacturers were more or less happy with their robots, and consequently industrial robots remained position controlled, somehow difficult to program by regular operators, and really not especially exciting machines. Features currently common in research laboratories hadn't reached industry yet because of some lack of interest from the robot manufacturing industry. Nevertheless, there was a considerable evolution that can be summarized as follows.

In 1974 the first electrical drive trains were available to use as actuators for robot joints. In the same year the first microprocessor controlled robots were also available commercially.

Around 1982, things like Cartesian interpolation for path planning were available in robot controllers, and many of them were also capable of communicating with other computer systems using serial and parallel interfaces. In the same year, some manufacturers introduced joystick control, for easier programming, and *teach pendant* menu interface.

In 1984, vision guidance was introduced as a general feature being used for tracking, parts identification, *etc.*

In 1986, the first digital control loops were implemented enabling better actuator control and enabling the use of AC drives.

Networking is a feature of the 1990s with several manufacturers implementing networking capabilities and protocols.

In 1991 there was the implementation of digital torque control loops which enabled, for example, the utilization of full dynamical models, which was a feature available in the first robots around 1994.

During the period 1992-1994 several manufacturers introduced features like Windows-based graphical interfaces, virtual robot environments for off-line programming, and *fieldbuses*.

Robot cooperation is a feature introduced from 1995 to 1996.

Figure 1.7. Several current robot manipulators available on the market

Around 1998 robot manufacturers started introducing collision detection to avoid damaging robots, and load identification to optimize robot performance. Since then other features include fast pick and place, weight reduction, optimized programming languages, object oriented programming, remote interfaces using RPC sockets and TCP/IP sockets, *etc.*. Figure 1.7 shows some of the robot manipulators available currently on the market.

And how do we define robotics then? Is it a science? Is it a technique or collection of techniques? If the reader takes a robotics book then something like this appears:

> *"A robot is a re-programmable multi-functional manipulator designed to move materials, parts, tools, or specialized devices, through variable programmed motions for the performance of a variety of tasks", from the book Robotics – Control, Sensing, Vision and Intelligence, Fu, Gonzalez, Lee, MacGraw Hill, 1987.*

Although correct, despite being restricted to robot manipulators, this definition does not give the correct idea. The common-sense image of a robot is usually associated with strong and superb machines, tireless (like *Karel's Capek* machines), obedient ("*yes, noberto san ...*"), but nevertheless, fascinating machines that make us dream. And that fascination is not in that definition.

Like with everything, we should look to the past and pick what was fundamental for the history of robotics in terms of ideas and dreams. From the *Greeks* and *Arabs* we should pick the idea of "*ingenious devices*". In fact, robotics is very much about mechanics, motion, mechanisms to transmit motion, and having the art and the skill to design and build those mechanisms. Yes, "ingenious devices" is really a good start.

Then we should listen to *Leonardo* (sixteenth century) and look to his quest on "*... precision ...*" and "*...permanent power source ...*". He understood that robots need parts built with very high precision and a permanent power source. That was not available at his time, *i.e.*, machine tools and a permanent power source (electric, hydraulic or pneumatic).

Finally, we should read *Nicola Tesla* and observe his outstanding and visionary work. He understood after all that robots are a consequence of dreams and neat ideas. Robots need to be controlled and programmed, distinguish situations, *etc.*, have ways of "*understanding*", and that means computers, electronics, software, and sensors, in a way to enable machines to be programmed and to sense their environment. Those are the elements that enable us scientists, engineers, and robot users, to try different things and new ideas, being a source of fascination. In his own words [4]:

> *"... But this element I could easily embody in it by conveying to it my own intelligence, my own understanding. So this invention was evolved, and so a new art came into existence, for which the name "teleautomatics" has been suggested, which means the art of controlling movements and operations of distant automatons.*

Therefore, robotics is a science of generic ingenious mechatronical devices, precise, powered by a permanent power source and flexible, *i.e.*, open to new ideas and a stimulus to the imagination. A stimulus so strong that it attracted many of the best minds of our common history, *i.e.*, authors of the work that constitutes the legacy that we humans leave for the future.

1.3 Why to Automate Welding?

Robot manipulators came a long way since the early days. Actually robot manipulators are interesting machines in terms of flexibility, programmability and precision. Modern manufacturing systems depend increasingly on automatic equipments, namely robot manipulators. This is an economic choice based on the following reasons:

1. Robot manipulators can perform industrial tasks in a human-like manner with at least comparable quality for longer periods of time.
2. Robots manipulators present the best rate between "production cost" and "production volume" for small/medium production volumes (Figure 1.1). Actually that is the case of SMEs existing in developing and developed countries (as an example, SMEs represent more than 95% of the companies in Europe). In fact, considering actual market conditions (very high competition, products defined in part by the customers, products with low life cycles, increasing demand for higher quality at lower prices, *etc.*) companies operate based on orders and never risk big stocks (besides the necessary security stocks) which keeps production on small/medium scale.
3. Robot manipulators are unique flexible machines (mainly due to programmability) that can be adapted to perform very different tasks. Consequently, robot manipulators are suitable to be used with manufacturing setups requiring frequent task changes, which is fundamental to respond to market changes or to the introduction of new products.

Let's consider briefly an industrial solution developed recently (2004) by the first author. The objective is to justify the above-mentioned arguments just by introducing a system that takes advantage of robot manipulator capabilities and computer-based human-machine interfaces (HMI). The presented features are common to any high-demanding industrial system, namely systems requiring special human-machine interfaces and a semi-autonomous operation, like robotic welding.

1.3.1 Example of an SME Based Industrial Robotic System

The industrial robotic system presented in this section was designed to execute the task of removing the excess of joining PVC material from automobile glasses, which arises on its borders during the glass manufacturing cycle. In fact, most of the automobile glasses, namely front, rear and roof glasses, are composed of two sheets of glass joined by a layer of PVC. For proper assembly, and to ensure proper joining of the PVC to the glass while maintaining transparency, the glasses go through a heating process, followed by a considerable period inside a pressure chamber. This process generates a very stiff excess of PVC on the borders of the glass that must be carefully removed, since it alters the dimensions of the glass, causing difficulties in assembling it in the car body, and can have esthetical implications if for some reason the glass borders are not hidden.

Figure 1.8. Robotic glass deburring system

Traditionally this excess of PVC is removed by hand using small cutting devices. Nevertheless, for highly-efficient plants this is not desirable since it slows down production, and requires very high concentration from operators to avoid touching and damaging the glass with the cutting device. Consequently, the process is very risky for the quality of the final product. Furthermore, with recent car designs some glasses are glued directly in the chassis without any exterior rubber. This happens mainly with roof, front and rear glasses. Consequently, the requirements for perfect PVC removal are even higher, which demands an automatic procedure to execute it.

The system (Figure 1.8) designed to handle the operation described above is composed of [36]:

1. Two industrial robots ABB IRB6400 equipped with the S4C+ controllers.
2. Especially designed electric-pneumatic grippers to hold firmly the glasses.
3. Two automatic deburring belts controlled by the robot controller IO system.
4. One industrial PLC (Siemens S7-300) that manages the cell logic and the interface to the adjacent industrial systems, providing to the robot controllers the necessary state information and the interface to the factory facilities.
5. One personal computer to command, control and monitor the cell operation.

Briefly the system works as follows: the first robot verifies if conveyor 1 (Figure 1.8) is empty and loads it with a glass picked from the pallet in use. The system uses a rotating circular platform to hold three pallets of glasses, enabling operators to remove empty pallets and feed new ones without stopping production. After releasing the glass, the robot pre-positions to pick another glass which it does when the conveyor is again empty. If the working glass model requires deburring, then the centering device existing in the conveyor is commanded to center the glass so that the second robot could pick the glasses in the same position. With the glass firmly grasped, the deburring robot takes it to the deburring belts and extracts the excess of PVC by passing all the glass borders on the surface of the deburring belt. When the task is finished the robot delivers the glass on conveyor 2, and proceeds to pick another glass. The deburring velocity, pressure, trajectory, *etc.*, is stored in the robot system on a database sorted by the glass model, which makes it easy to handle several models. Programming a new model into the system is also very simple and executed by an authorized operator. There is a collection of routines that take the robot to pre-defined positions, adjusted by the given dimensions of the glass, allowing the operator to adjust and tune positions and trajectories. He can then "play" the complete definition and repeat the teaching procedure until the desired behavior is obtained. This means being able to control the robot operation with the controller in automatic mode, which is obtained by including some teach-pendant features in the process for operator interface.

Another important feature included in this robotic system is the possibility to adjust production on-line, adapting to production variations. This objective is obtained by using a client-server architecture, which uses the cell computer (client) to parameterize the software running on the robot controller (server). That can be achieved offering the following services from the robot server to the clients:

1. All planned system functionalities by means of general routines, callable from the remote client using variables that can be accessed remotely.
2. Variable access services that can be used remotely to adjust and parameterize the operation of the robotic system.

With this features implemented and with a carefully designed operator interface (Figure 1.9 and Figure 1.10) and robot server software, it's possible to achieve a system that requires limited human intervention related with adjustment tasks to cope with production variations. Since a remote interface is used (Figure 1.9 and Figure 1.10), the necessary adjustments are executed on-line without stopping production. Those operations include:

1. Adjusting the deburring angle, *i.e.*, the angle between the border of the glass and the deburring belt. The angle introduced is added to the programmed one, so that zero degrees means keeping the programmed angle unchanged.
2. Adjusting the force on the belt during the deburring operation (adjusted by position). The commanded value is entered in millimeters and updates the actual position in the direction perpendicular to the belt and parallel to the surface of the glass.
3. Adjusting the deburring speed.
4. Maintenance procedures necessary to change the belts after the planned deburring cycles.

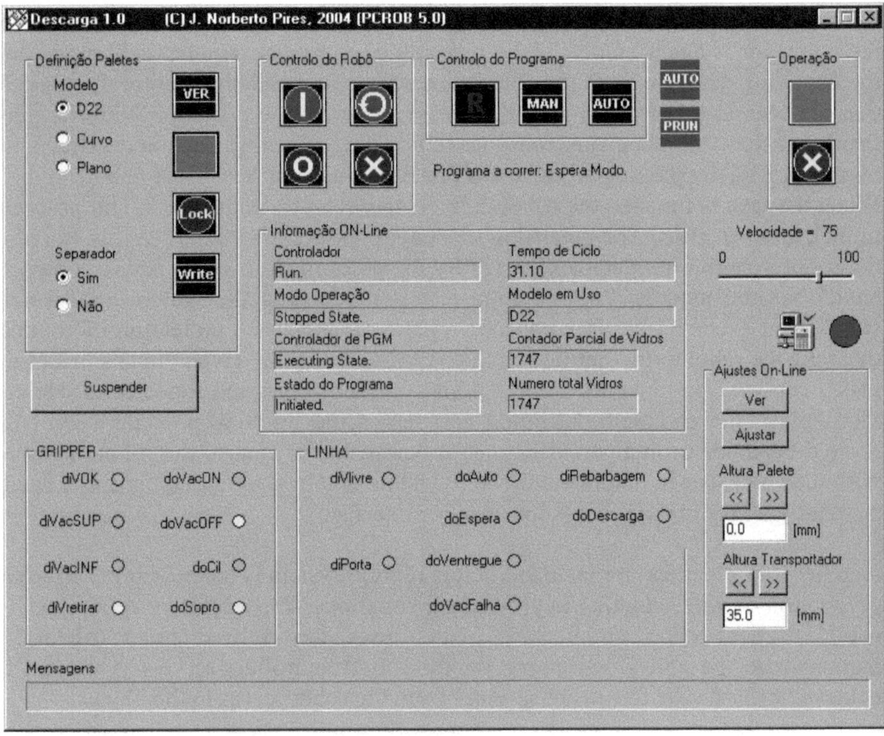

Figure 1.9. Operator interface for de-palletizing robot

The de-palletizing robot requires less parameterization since it executes a very simple operation. Besides that, the gripper adapts to the surface of every model of glass, using presence sensors strategically placed near two suction caps (see Figure 1.8), with the objective of having an efficient de-palletizing operation. Nevertheless, the operator is able to change the velocity of the process by stating a slow, fast or very fast cycle to adjust to production needs, suspend and resume operations, adjust the way the robot approaches the surface of the glass, *etc.*. These adjustments are necessary to obtain the most efficient operation in accordance with the observed production conditions, to solve daily problems and to cope with production variations.

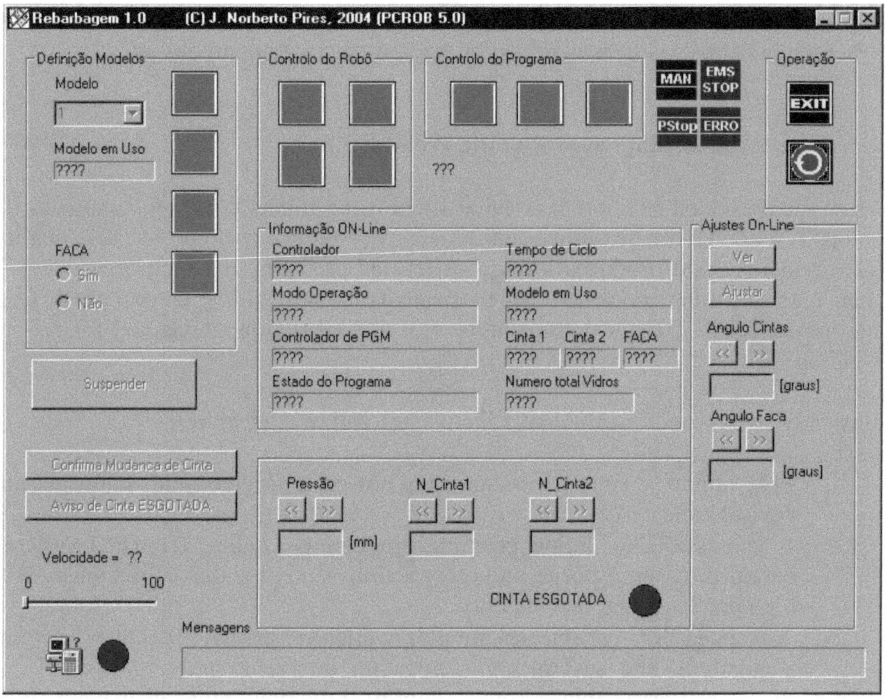

Figure 1.10. Operator interface for deburring robot

Finally, it is important to mention that the robot is equipped with a force/torque sensor mounted on the wrist. The objective is to adjust automatically the model setup introduced by the operator, correcting the points where the measured force between the belt and the glass exceeds the recommended values, attempting to avoid damage to the glass and to increase the deburring efficiency. This procedure is active during the process of applying a new model, and also during production, if explicitly activated by the operator, constituting an automatic correcting feature.

The system has worked for some time and proved to be very simple to operate with, showing also quick adaptation from operators [36]. The adjusting features added to the system proved to be very helpful, allowing the company to respond in a timely fashion to production changes, avoiding variations in the quality of the final product, and to introduce quickly new models into the production database. Since the models are identified automatically, using barcode readers placed on the pallet platform, the system works continuously without operator intervention. The only thing needed is to feed the system with pallets full of glasses, removing the empty ones. That operation is done periodically with the help of electro-mechanical fork-lift-trucks.

Most of the features presented for this example will be explored in this book for robotic welding applications, namely the capacity to simulate the procedure, the capacity to adjust on-line and change parameterization, the capacity to monitor the system, the database like way of specifying sequence of operations, *etc.*

1.3.2 Are Robots Adapted to Robotic Welding?

From the presented example it is evident that using robots in actual manufacturing setups is a choice for flexibility, agility, a way to reduce cost and increase quality. The modern world produces a huge variety of products that include welding in their manufacturing processes, which means that also welding could benefit from the introduction of robot manipulators. But are actual robots adapted for robotic welding?

Basically actual robot manipulators include the following features:

1. Programmable control system, using powerful programming languages and environments.
2. It is possible to define positions/orientations, define reference systems, parameterize trajectories and other actions, and play that continuously with high precision and repeatability.
3. Advanced PLC capabilities are also available, namely, IO control and data acquisition, and several communication interfaces and protocols. These functionalities enable robots to coordinate actions with other equipments and sensors, and being integrated with other computers and manufacturing systems existing in the setup.

The most important characteristics of actual robot manipulators are summarized in Table 1.1.

Since most welding techniques require motion control, sensor integration and coordination with the welding power source (controlled using IO digital and analog signals, or *fieldbuses*), then robot manipulators are an almost perfect match for the vast majority of welding processes. Difficulties may also arise in automating the welding process, namely using robots. In fact, introducing robots means increasing complexity in the manufacturing process, and requires skilled personnel to handle programming and maintenance. That may constitute a major drawback, not allowing companies to take full advantage from the flexibility stored inside the robotic manufacturing machines. This puts focus on human-machine interfaces (HMI) for control, command and supervision, leaving space for the software architecture used to develop the HMI solutions.

In conclusion, the majority of industrial welding applications benefit from the introduction of robot manipulators, since most of the deficiencies, attributed to the human factor, are removed with advantages when robots are introduced. Also, the welding process is very dangerous and demanding in precision and operator attention, requiring substantial physical efforts from operators, which makes it a good candidate for robots.

Table 1.1. Robot manipulators main characteristics.

Repeatability	Up to 0.03 mm (0.1 mm is common)
Velocity	Up to 5 m/s
Acceleration	Up to 25 m/s^2
Payload	From around 2-3kg up to ~750 kg
Weight/Payload	Around 30-40
Axis	6
Communications	*Profibus*, *can*, *devicenet*, *ethernet* and serial channels (RS232 and RS485)
IO Capabilities	PLC like capabilities to handle digital and analog IO

1.4 Objectives and Outline of the Book

The present book gives a detailed overview of Robotic Welding in the beginning of the twenty-first century. The evolution of robotic welding is presented, showing to the reader what were the biggest steps and developments observed in the last few years. This is presented with the objective of establishing the current *state-of-the-art* in terms of technologies, welding systems, software and sensors. The remaining issues, *i.e.*, the issues that remain open are stated clearly, in a way to motivate the readers to follow the rest of the book which will make contributions to clarify most of them and help to solve a few.

To do that, a chapter on "*Welding Technology*" is presented, describing the most important welding techniques and their potential and requirements for automation using robot manipulators. This chapter includes established results on robotic welding processes, which can constitute a good source of information for readers and also good source of examples.

Also, a revision with current research results on "*Sensors for Welding Robots*" used on robotic welding is presented in the book. That includes sensors for seam tracking, quality control and supervision. This chapter includes all system requirements necessary to use those sensors and sensing techniques with actual robot control systems. Hardware and software interfaces are also covered in detail.

A revision of available welding systems, including hardware and software, clarifying their advantages, and drawbacks is also presented to give to the reader a clear picture of the area. This is included in the chapter "*Robotic Welding: System Issues*".

Finally, a few industrial applications using the presented techniques and systems are presented. The book includes a chapter on "*Robotic Welding: Application Examples*", where a few selected applications are described in detail including aspects related to software, hardware, system integration and industrial exploitation. This chapter uses actual robots, but it is presented in a general way so that the interested reader can easily explore his own interest.

A good collection of references is also presented at the end of each chapter, to enable the reader to explore further from the literature.

1.5 References

[1] M. Rosheim, "Robot Evolution: The Development of Anthrobots", New York: John Willey & Sons, 1994.

[2] M. Rosheim, "In the Footsteps of Leonardo", IEEE Robotics and Automation Magazine, June 1997.

[3] C. Pedretti, "Leonardo Architect", Rizzoli International Publications, New York, 1981.

[4] N. Tesla, "My Inventions: Autobiography of Nicola Tesla", Willinston, VT: Hart Brothers, 1983.

[5] Myhr, M., "Industrial New Trends: ABB view of the Future", International Workshop on Industrial Robotics, New Trends and Perspectives (http://robotics.dem.uc.pt/ir99/), Parque das Nações, Lisbon, 1999.

[6] United Nations and International Federation of Robots, "World Industrial Robots 1996: Statistics and Forecasts", New York: ONU/IFR, 2000.

[7] Kusiak A., "Modelling and Design of Flexible Manufacturing Systems", Elsevier Science Publishers, 1986.

[8] Kusiak A., "Computational Intelligence", John Wiley & Sons, 2000.

[9] Bolmsjo G., "Sensor System in Arc Welding", Technical Report, Lund Institute of Technology, Production and Materials Engineering Department, 1997.

[10] Bolmsjö G., Olsson M., Nikoleris G., and Brink K. Task programming of welding robots. In Proceedings of the Int. Conf. on the Joining of Materials, JOM-7, pages 573-585, May-June 1995.

[11] Loureiro A., Velindro M. and Neves F., "The influence of Heat Input and the Torch Weaving Movement on Robotized MIG Weld Shape", International Journal for the Joining of Material, Vol.10, N.3/4, pp. 86-91,1998.

[12] Agren B. "Sensor Integration for Robotic Arc Welding." PhD thesis, Lund University, 1995

[13] Richardson RW, "Robotic Weld Joint Tracking Systems - Theory and Implementation Methods", Welding Journal, Nov. 1986.

[14] Books B., "Welding Robots - The State of the Art", Welding and Metal Fabrication, June 1991.

[15] Drews P., Starke G., "Development Approaches for Advanced Adaptive Control in Automated Arc Welding", Internal Report XII-970, Mechatronics Department, Achen, Germany, 1986.

[16] Yada Y. *et al*, "Development of Real Time Welding Control System by Using Image Processing", Internal Report XII-998, Mechatronics Department, Achen, Germany, 1986.

[17] Dahlén P. and Bolmsjö G., "Human Factors in the Justification of Advanced Manufacturing Systems". International Journal of Human Factors in Manufacturing, 6(2):147-162, 1996.
[18] Pires JN, Sá da Costa JMG, "Object Oriented and Distributed Approach for Programming Robotic Manufacturing Cells", IFAC Journal on Robotics and Computer Integrated Manufacturing, February 2000.
[19] Pires, JN, Monteiro, P, Schölzke, V, "Using Robot Manipulators on High Efficient Wrapping Machines for Paper Industry", ISR'2001, Seoul, Korea, April 2001.
[20] Pires, JN, "Using Matlab to Interface Industrial Robotic & Automation Equipment", IEEE Robotics and Automation Magazine, September 2000.
[21] Sá da Costa, JMG, Pires, JN, "Future Welding Robot Developments", Journal Robótica, n°41, January 2001.
[22] Pires, JN, "Programming Industrial Robotic and Automation Equipment", Industrial Robot, An International Journal, MCB University Press, July 2000.
[23] Touret, S., "Le procédé TIME", Soudage et Techniques Connexes, pp. 29-33, May-June, 1997.
[24] Choi, S.K., Yoo C.D. and Kim Y.S., "Dynamic Simulation of Metal Transfer in GMAW. Part1: Globular and Spray Transfer Mode Welding", pp. 38-44, "Part2: Short Circuit Transfer Mode", pp. 45-51, Welding Journal, 77 (1), 1998.
[25] Jones, LA Edgar, TW and Lang JH, "Images of Steeel Electrode in Ar-20% Shielding During Constant Current Gas Metal Arc Welding", Welding Journal, 77 (1), pp. 135-141, 1998.
[26] The Welding Institute, "Standard Data for Arc Welding", Abington Hall, Abington, UK.
[27] Adolfsson S., Bahrami A., Bolmsjö G. and Claesson I.. "On-Line Quality Monitoring in Short-Circuit Gas Metal Arc Welding." Welding Journal, 75:59s-73s, February, 1999
[28] Bolmsjö G., "Knowledge Based Systems in Robotized Arc Welding. Knowledge Based Systems - Advanced Concepts", Techniques and Applications, (Spyros Tzafestas, editor). Chapter 17, pp. 465-495. World Scientific, ISBN 9810228309, 1997.
[29] Proceedings of the International Workshop on Robotic Welding Systems and Process Monitoring (http://robotics.dem.uc.pt/welding/), Coimbra Portugal, January 2001, published at the Journal Robótica, January 2001.
[30] Meta Systems (UK), http://www.meta.co.uk
[31] Servo-Robot (Canada), http://www.servorobot.com
[32] ABB IRB1400 Users Manual, ABB Flexible Automation, 1997.
[33] RAP, Service Protocol Definition, ABB Flexible Automation, 1996.
[34] Bloomer J., "Power Programming with RPC", O'Reilly & Associates, Inc., 1992.
[35] Smith, C.S., Gnudi M.T., "The PIROTECHNIA of Vannoccio Biringuccio, The classic Sixteenth-Century Treatise on Metals and Metallurgy", Dover Publications Inc., New York, 1990.
[36] Pires JN, "Semi-autonomous Manufacturing Systems: the role of the HMI software and of the manufacturing tracking software", IFAC Journal on Mechatronics, accepted for publication on Vol. 15, to appear in 2005.

2

Welding Technology

2.1 Gas Tungsten Arc Welding (GTAW)

The gas tungsten arc welding (GTAW) process is based on the electric arc established between a non-consumable electrode of tungsten and the work-pieces to be joined. Part of the heat generated by the electric arc is added to the work-pieces, promoting the formation of a weld pool. The weld pool is protected from air contamination by a stream of an inert gas (Ar or He) or a mixture of gases.

2.1.1 Introduction

This process is also known as tungsten inert gas (TIG), although small amounts of non-inert gases may be used in the shielding mixture, such as hydrogen or nitrogen. Figure 2.1 illustrates the principal elements of the conventional process.

Autogenous GTAW welding (without filler metal) is used in thin square edged sections (2mm), while V and X type edge preparations are needed in thicker sections. In this case, the addition of filler metal is necessary. This process is extensively used for welding thin components of stainless steel, aluminum, magnesium or titanium alloys as well pieces of carbon and low alloy steels [1],[2].

Heat input in GTAW does not depend on the filler material rate. Consequently, the process allows a precise control of heat addiction and the production of superior quality welds, with low distortion and free of spatter. It is less economical than other consumable electrode arc welding processes, due to its lower deposition rate, and it is sensitive to windy environment because of the difficulty in shielding the weld pool. Besides it shows low tolerance to contaminants on filler or base metals.

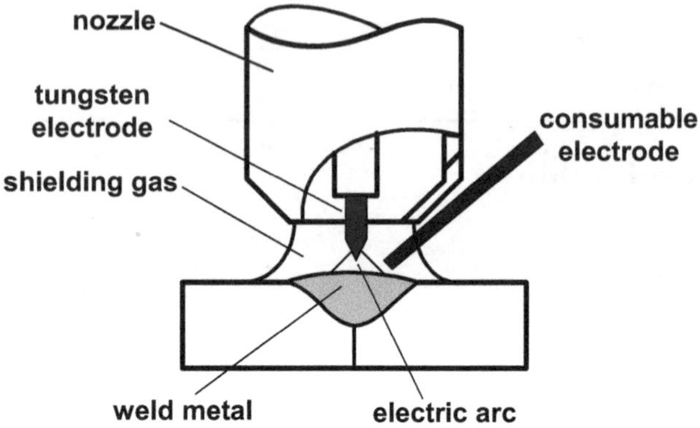

Figure 2.1. Diagrammatic sketch of the gas tungsten arc welding process (GATW)

The autogeneous process is readily used in robotics, although special techniques are needed when it is necessary to add filler metal to the weld pool.

2.1.2 Welding Equipment

In this section the relevant aspects related to the welding equipment used with the GATW process will be reviewed, with the objective of exploring the implications for automatic robotic welding.

2.1.2.1 Power Sources

Power sources for GTAW are generally of the constant current type with drooping volt-ampere static curves, as illustrated schematically in Figure 2.2. Light weight transistorized direct current power sources are currently used, being more stable and versatile than the old thyristor-controlled units [3]. In rectifier-inverter power sources the incoming AC current is rectified and then converted into AC current at a higher frequency than that of the mains supply, in the inverter. Afterwards high voltage AC current is transformed into low voltage AC current suitable for welding, in the transformer, and then rectified, as shown schematically in Figure 2.3. The aim to increase the current frequency is to reduce the weight of the transformer and other components of the source such as inductors and capacitors.

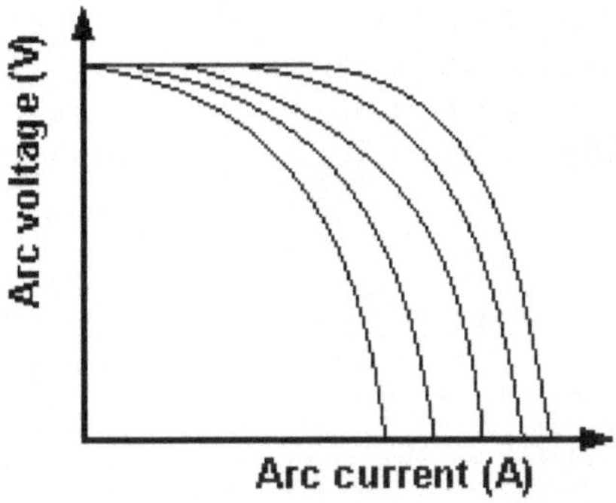

Figure 2.2. Plot of the arc voltage *vs* current voltage for GTAW power sources

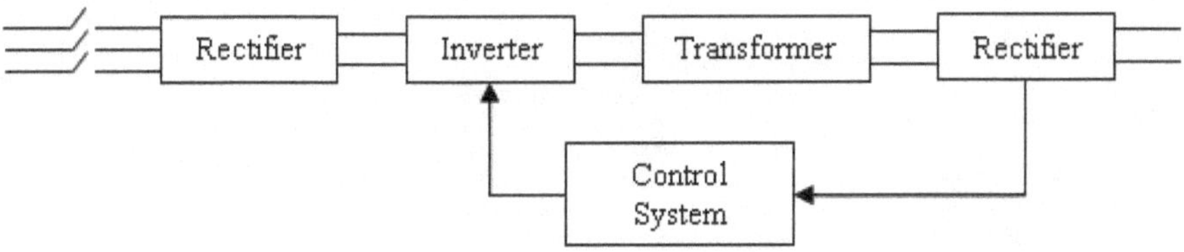

Figure 2.3. Sketch of the inverter principle of the power sources

2.1.2.2 Welding Torch

The welding torch holds the non-consumable electrode, assures the transfer of current to the electrode and the flow of shielding gas to the weld pool. Torches with welding regimes up to 200 A are generally gas-cooled and those with continuous operation between 200 and 500 A are water-cooled. Figure 2.4 shows an exploded view of a water-cooled torch.

2.1.2.3 Non-consumable Electrodes

Non-consumable electrodes are composed of pure tungsten or of tungsten alloys. Pure tungsten electrodes can be used with DC but are more sensitive to contamination, have lower service life-cycle and exhibit higher tip deterioration than alloyed electrodes. These electrodes can be used in welding of aluminum and magnesium alloys on AC.

Thoriated tungsten (2% ThO_2) electrodes are widely used in industrial applications due to its excellent resistance to contamination, easy arc starting and stable electric arc. Concerns about safety, because thorium oxide is radioactive, led to the development of other electrodes containing small proportions (around 2%) of simple earth rare elements such as lanthanum, yttrium and cerium or even mixtures of several elements [4],[5]. These electrodes have better operational characteristics than thoriated electrodes and can be used in welding carbon and stainless steels, nickel and titanium alloys. Zirconiated tungsten electrodes are excellent for AC due to its good arc starting, high resistance to contamination and small tip shape deterioration.

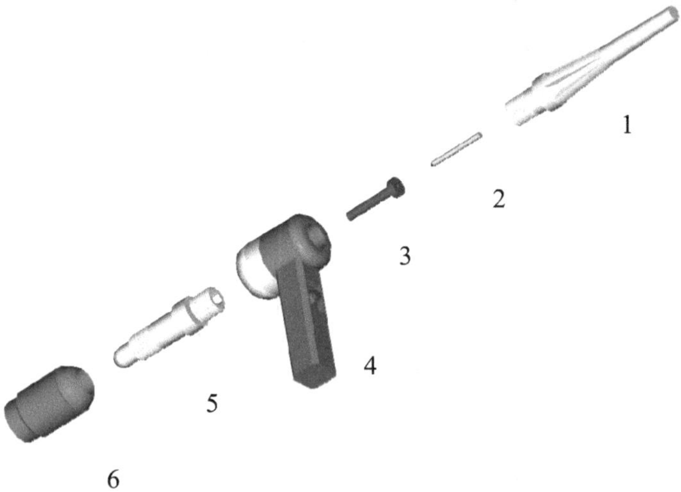

Figure 2.4. Exploded view of a torch: back cap – 1; electrode – 2; collet – 3; handle – 4; collet body – 5; nozzle – 6

These electrodes are available in diameters between 0.5 and 12 mm, although the most usual are up to 4 mm in diameter, being the normal length between 50 and 175 mm. The selection of the electrode diameter to use depends on the plate thickness to be welded, being in general similar to plate thickness.

2.1.2.4 Arc Striking Techniques

Arc initiation by touch striking was used formerly in manual GTAW, but this technique is very sensitive to tungsten contamination, adversely affecting the service life of the electrode. High-frequency-high-voltage (*e.g.* 3 kV at 5 MHz) supplies are currently used in arc striking and AC arc stabilization in manual GTAW systems [3]. This arc starting technique usually produces interference in electronic equipment in the vicinity of the power source.

Programmed touch striking is an alternative technique developed for automatic systems. In this technique current and voltage are limited when electrode touches in the work-piece, in order to prevent electrode contamination. A pilot arc starting can also be used to initiate the main electric arc, though a more complex torch is needed.

2.1.2.5 Shielding Gas Regulator

The regulator is a device that reduces source gas pressure to a constant working pressure, independently of source pressure variations. Pressure reduction can be made in one or two stages. Regulators in two stages give in general more stable output flow.

2.1.3 Process Parameters

In this section the relevant parameters for the GTAW process will be reviewed with the double objective of presenting them and showing that they can certainly be used for automatic robotic welding.

2.1.3.1 Current

Current has direct influence on weld bead shape, on welding speed and quality of the weld. Most GTAW welds employ direct current on electrode negative (DCEN) (straight polarity) because it produces higher weld penetration depth and higher travel speed than on electrode positive (DCEP) (reverse polarity). Besides, reverse polarity produces rapid heating and degradation of the electrode tip, because anode is more heated than cathode in gas tungsten electric arc.

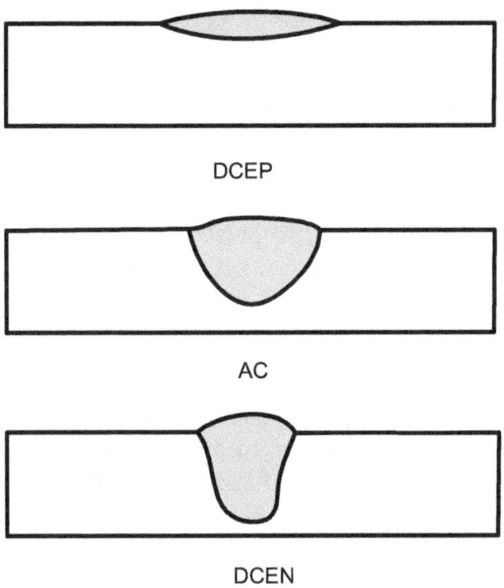

Figure 2.5. Effect of current and polarity on weld bead shape

Reverse polarity may be of interest in welding aluminum alloys because of the cathodic cleaning action of negative pole in the work-piece, that is the removal of the refractory aluminum oxide layer. However alternating current is better adapted to welding of aluminum and magnesium alloys, because it allows balancing electrode heating and work-piece cleaning effects. Weld penetration depth obtained with AC is between depth obtained with DCEN and DCEP, as illustrated in Figure 2.5.

Square wave AC is nowadays being used instead of the normal sine wave because it facilitates the assistance of the arc re-strike each half cycle and allows adjusting of the arc cleaning effect or the penetration depth. Cleaning action is improved by increasing duration of the electrode positive half cycle. The increase in penetration depth is given by increasing the duration of the electrode negative half cycle, as shown schematically in Figure 2.6.

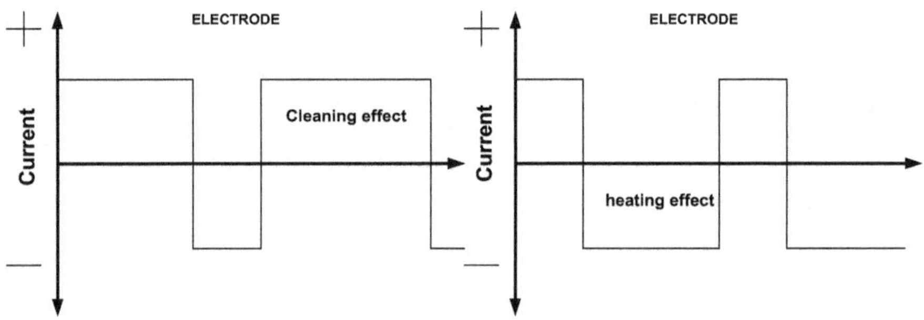

Figure 2.6. Influence of the balance between alternate half cycles on GTAW

Pulsed DC current with low-frequency (1-10 Hz) is being used to reduce weld distortion, to improve tolerance to joint preparation and to cast-to-cast variations. Current magnitude and duration of the pulses are determined by material family and thickness of the component to be welded and are related by Equation 2.1 [3],

$$I_p \cdot t_p = K \qquad 2.1$$

where, I_p is the pulse current, t_p is the pulse time and K is a constant. Background current and time are selected in order to allow solidification of metal between pulses. This current is used in welding of stainless steels.

High-frequency pulsed current (5-30 kHz) improves arc stiffness, increasing penetration depth and maximum welding speed and decreasing formation of porosity in the weld metal. This current is advantageous in automatic welding applications.

2.1.3.2 *Welding Speed*

The effect of increasing the welding speed for the same current and voltage is to reduce the heat input. The welding speed does not influence the electromagnetic force and the arc pressure because they are dependent on the current. The weld speed increase produces a decrease in the weld cross section area, and consequently penetration depth (D) and weld width (W) also decrease, but the D/W ratio has a weak dependence on travel speed [7]. These results suggest that the travel speed does not influence the mechanisms involved in the weld pool formation, it only influences the volume of melted material. Normal welding speeds are from 100 to 500 mm/min depending on current, material type and plate thickness.

2.1.3.3 *Arc Length*

The arc length is the distance between the electrode tip and the work-piece. The arc length in GTAW is usually from 2 to 5 mm. If the arc length increases, the voltage to maintain the arc stability must increase, but the heat input to work-piece decreases due to radiation losses from the column of the arc. Consequently, weld penetration and cross section area of melted material decrease with increasing arc length.

2.1.3.4 *Shielding Gases*

Shielding gases are used in GTAW in order to prevent atmospheric contamination of the weld metal. This contamination can produce porosity, weld cracking, scaling and even change in the chemical composition of melted material. Besides shielding gas also has a large influence on the stability of the electric arc. Gases with low ionization potential facilitate the ignition of the electric arc and those with low thermal conductivity tend to increase the arc stability.

Argon is the most used GTAW shielding gas. It has low ionization potential and is heavier than air, providing an excellent shielding of the molten weld pool. Furthermore it is less expensive than helium, the other inert shielding gas used in the process. Argon is used in welding of carbon and stainless steels and low thickness aluminum alloys components.

For welding thick aluminum work-pieces and other high-conductive materials, such as copper alloys, helium is recommended because it has higher ionization potential than argon, needing higher voltage for arc initiation and maintenance, but producing higher heat-input. Helium or helium/argon (30-80% He) mixtures allow increased welding speed and improved process tolerance.

Mixtures of argon with up to 5% of hydrogen are frequently used in welding of austenitic stainless steels. Hydrogen increases arc-voltage and consequently heat-input, increasing weld penetration and weld travel speed, as well improving weld appearance [6]. Argon/hydrogen mixtures are also used in welding of copper-nickel alloys.

Argon is also used as back side shielding gas, mainly in welding of stainless steels, aluminum alloys and reactive metals.

Flow rates of shielding gases depend on weld thickness, being 4-10 l/min for argon and 10-15 l/min for helium, because it is lighter than argon, and consequently less effective in shielding.

Gases with a purity of 99.995% are used in welding most of the metals, though reactive materials such as titanium need contaminant level less than 50 ppm.

2.1.3.5 Filler Metals

Filler metals are generally used for plate thickness above 2 mm, having chemical composition similar to that of the parent material. Filler metal diameter is between 1.6 and 3.2 mm and in automatic systems is normally added cold from a roll or a coil.

2.1.3.6 Electrode Vertex Angle

The non-consumable electrode angle influences the weld penetration depth and the weld shape [7]. Electrode angles between 30° and 120° are used. Small angles increase arc pressure and penetration depth but have high tip shape deterioration. Electrode angles from 60° to 120° maintain tip shape for longer periods and give welds with adequate penetration depth-to-width ratio.

2.1.3.7 Cast-to-cast Variation

Cast-to-cast variation refers to variation observed in penetration of welds produced in the same welding conditions in several batches of austenitic stainless steel with nominally identical composition. These changes in the weld bead shape are attributed to variation in proportion of trace elements in the material, such as sulphur, calcium and oxygen. Variations in trace elements seem to affect surface tension and metal flow into the pool [8]. Weld pool shape is also affected by electromagnetic forces, arc pressure and thermo capillarity forces [9]. To minimize this problem several strategies have been adopted such as the use of higher currents or of pulsed current, the application of adequate shielding gases or the application on plate surface of flux coatings containing active ingredients [10].

2.1.4 Process Variants

GTAW is regarded as a high quality process for welding thin metals using low travel speed and low electrode deposition rate, requiring highly skilled personnel in manual welding. Variants developed seek to improve productivity, mainly deposition rate, penetration depth and welding speed. These variants are implemented in automatic or robotic systems.

Hot-wire GTAW is a variant where a heated filler wire is fed to the rear of the melted weld pool at a constant rate, as represented schematically in Figure 2.7. Filler wire is resistance heated close to melting point using mainly AC power sources, in order to minimize magnetic disturbance of the electric arc. Deposition rates up to 14 kg/h can be attained with this process. It has been used in heavy wall fabrication, maintaining high joint integrity [11].

The use of a dual-shielding GTAW technique, see Figure 2.8, where an additional concentric gas shield gives an increase in constriction and stiffness of the electric arc, may be used to increase welding speed and penetration depth [12]. Constriction of the arc is produced by the external cold gas flow which decreases temperature of the outer part of the arc, decreasing the arc cross section where current flow occurs, consequently increasing current density and temperature. Electrode gas and annular gas may be of the same or of different compositions, such as Argon plus 5% hydrogen for internal gas and argon for external gas when welding austenitic stainless steels. For currents above 335A keyhole welding is obtained and the process may becomes sensitive to the process parameters. This technique also tends to increase the risk of undercut [3].

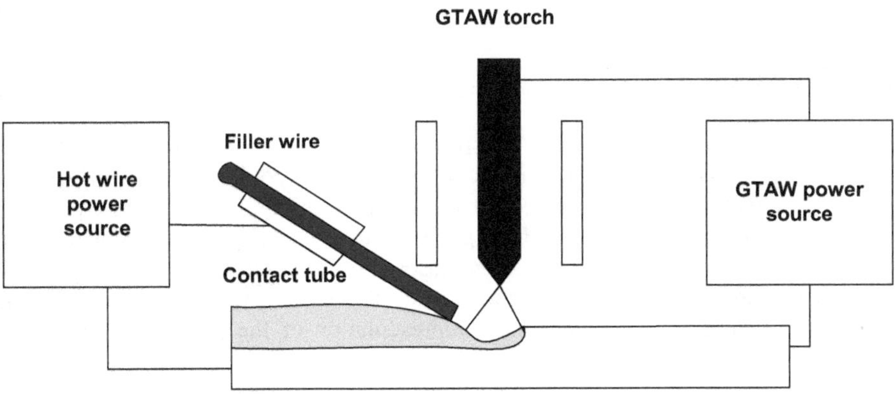

Figure 2.7. Schematic representation of a GTAW hot wire system

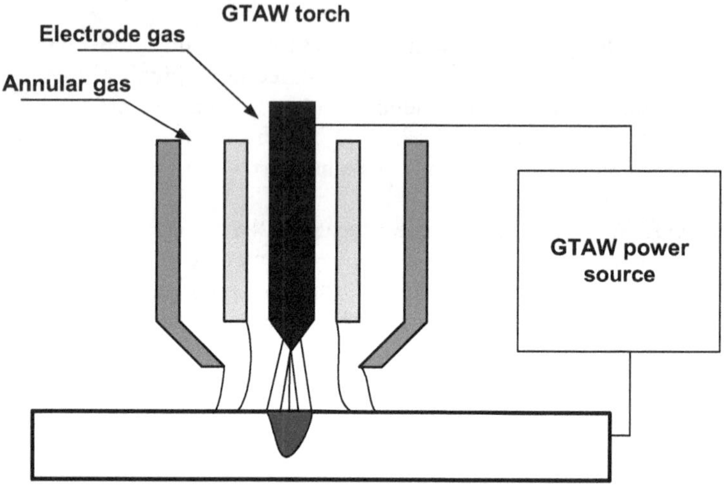

Figure 2.8. Schematic representation of dual-shielding GTAW system

Very high currents (I > 300 A) may also be used in a conventional automated GTAW process to increase the penetration depth, but defects may form and the process becomes unstable above 500 A. The keyhole mode gas tungsten arc welding process, which was developed a few years ago, seems to be suitable for ferrous and non-ferrous materials in the range from 3 to 12 mm [13]. However, this keyhole technique is extremely sensitive to arc voltage, and loss of material may occur through the keyhole vent.

2.2 Gas Metal Arc Welding (GMAW)

In the gas-metal arc welding (GMAW) process n electric arc is established between a consumable electrode, fed continuously to the weld pool, and the work-piece. Initially the weld pool was shielded by an inert gas, giving the process the popular designation of metal inert gas (MIG). Nowadays active gases such as carbon dioxide or mixtures of inert and active gases are also used and metal-active gas (MAG) is a common process nomenclature in this case. The designation GMAW includes all these cases. A schematic representation of the process is shown in Figure 2.9.

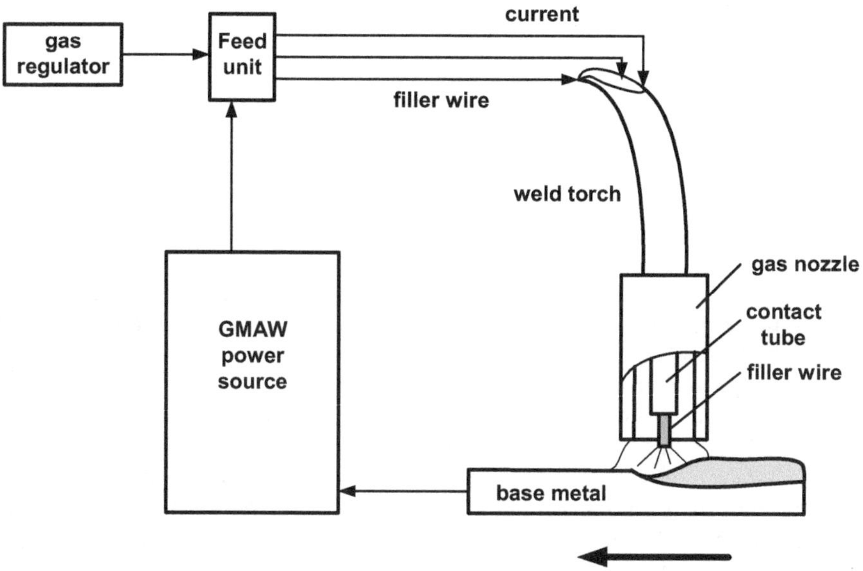

Figure 2.9. Schematic representation of gas metal arc welding process (GMAW)

2.2.1 Introduction

This process is widely used in industrial applications due to its numerous benefits. It can weld almost all metallic materials, in a large range of thicknesses (above 1 mm up to 30 mm or more) and is effective in all positions. GMAW is a very economic process because it has higher speeds and higher deposition rates than for example the manual metal arc process, and does not require frequent stops to change electrodes, as is the case of this former process. In addition, minimal post weld cleaning is needed because slag is almost absent. Less operator skill is required than for other conventional processes because electrode wire is fed automatically (semi-automatic process) and a self-adjustment mechanism maintains the arc length approximately constant even when the distance weld torch to work-piece varies within certain limits. These advantages make the process very well adapted to be automated and particularly to robotic welding applications.

The process is sensitive to the effects of wind, which can disperse the shielding gas, and it is difficult to use in narrow spaces due to the torch size. Problems such as lack of shielding, irregular wire feeding, unstable arc, burn-back or even weld discontinuities (porosity, incomplete penetration, excessive melt-through, undercutting or cracks) can occur during welding [14].

2.2.2 Welding Equipment

Basic equipment for conventional GMAW is consists of the power source, the electrode feed unit, the welding torch and the shielding gas regulator, as represented schematically in Figure 2.9.

2.2.2.1 Power Source

Most common GMAW power sources are of the inverter type with an architecture similar to that represented in Figure 2.3, but providing a constant-voltage output. A constant-voltage power source used in conjunction with a constant speed wire feeder can provide self-adjustment and stabilization of the arc length, in order to compensate for the variations in the torch to work-piece distance that occur mainly during manual welding operations. In a power source with approximately constant-voltage characteristics any change in the arc length is compensated by the modification of the weld current and consequently of the burn-off behavior of the electrode. Figure 2.10 illustrates the effect of increasing the arc length from L_1 to $L2$, which corresponds to an increase of the torch to work-piece distance. This increase of arc length produces an increase of the arc voltage and consequently a decrease of the weld current from I_1 to I_2 and of the burn-off rate from B_1 to B_2. As the wire feed speed is constant and burn-off decreases the arc tends to assume the initial length.

In addition these machines provide slope control of the power source characteristics and of the inductance in order to control spatter in short-circuiting transfer [3]. Inductances introduced in the output circuit reduce the rate of rise of current during the short-circuiting, reducing in this way the risk of explosion of metal droplets. In the case of thicker electrodes, which show a small variation of burn-off rate with current, or for materials having high conductivity, such as aluminum, process control is achieved by using a variable-speed wire feed unit that reacts to the arc length changes by adjusting the electrode feed speed.

GMAW inverters are also used to generate pulsed current with pulsed repetition rates (PRR) (number of pulses per second) typically between 100 and 200 PRRs [15]. Pulsed parameters are defined by algorithms in the controller. New synergic pulsed GMAW inverters can control melting rate through the modulation of the pulse shape and of the pulse frequency, being the process managed by a microprocessor [3].

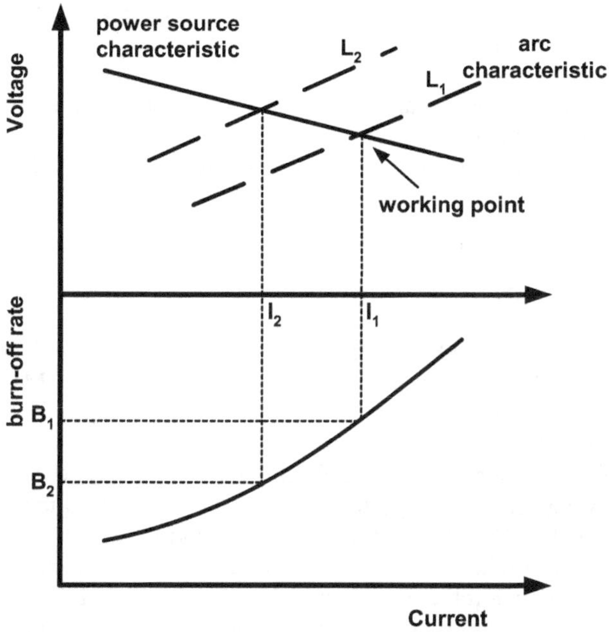

Figure 2.10. Self-adjustment mechanism with a constant-voltage power source. Arc length $L_1 > L_2$

2.2.2.2 Electrode Feed Unit

The electrode feed unit and the welding control mechanism are generally furnished in one integrated package. The electrode feed unit pulls the electrode from the reel and pushes it through a conduit to the welding torch (gun). This unit is composed of a direct-current motor, that varies the motor speed over a large range, a gear box and two pairs of rolls with a pressure adjusting screw and wire guides, that transmit mechanical energy , straighten and guide the electrode. Knurled rolls are used for hard materials, such as steel electrodes, and V and U type rolls are used for softer materials, such as aluminum electrodes. For soft electrodes or long conduits push-pull systems can be used too. Theses systems are composed of two feed units, one that is close to the wire reel that pushes the electrode, and the other unit in the torch that pulls the electrode. In automatic and robotic welding systems the electrode is fed from a spool (15-18 kg) or large drum (200-475 kg) to minimize wire supply changing. Normally, the electrode feeder for robotic welding is mounted separate from the power supply.

The welding control mechanism regulates not only the electrode feed speed and the start and stop of the electrode but also the delivery of shielding gas, current and cooling water (when necessary) to the torch. Creep start, gas pre-flow and post-flow, hot start, crater filling and adjustable burn-back time can frequently be programmed in this unit. Memory for pre-programs and for set parameters is frequently available in this unit.

When the torch cable is externally attached to the robot arm it is exposed to work-piece interference and to premature wear. Modern robotic systems can include special arms with internal cabling, in order to prevent interference, increasing cable life.

2.2.2.3 Welding Torch

Main functions of the welding torch are to furnish the electrode with electrical current and direct the electrode and gas flow to the work-piece. Main components of the welding torch are the contact tube, where the current is transmitted to the electrode, the nozzle, which provides a laminar gas flow to the weld pool, the torch switch, which sends signals to the feed unit, and the handle. The handle supports the gas and water (if necessary) tubes, the electrode guide tube and cables for current and signals. MIG torches for low current and light duty cycle (up to 60%) are gas cooled and torches for heavy duty cycle (up to 100%) and high current are water cooled. Robotic torches are in general water cooled, but if gas cooled torches are used they must be larger than manual torches. Alternatively air cooled torches, which use shop compressed air, can be applied instead of water cooled torches [16]. Robotic torches usually have emergency-stop capability to prevent damage to the robot arm and the welding torch in the event of a collision. They are also provided with automatic cleaning, that may include a pressurized air system for blowing spatter out of the nozzle, a reamer for cleaning the internal nozzle structure and an anti-spatter fluid delivery system.

Twin-wire GMA robotic welding torches can be used to reach higher deposition rate and welding speed. In this case a side-by-side configuration is used, with both wires being fed to close contact tips, in order to give a single weld pool.

2.2.3 Process Parameters

Welding parameters affect the way the electrode is transferred to the work-piece, the arc stability, spatter generation, weld bead geometry and overall weld quality. The main parameters of the process are current, voltage, travel speed, electrode extension and electrode diameter, though others, such as electrode orientation, electrode composition and shielding gas, also have direct influence on the metal transfer mechanisms. These parameters are not independent. The current and voltage, for example, are correlated by the arc characteristic curves shown in Figure 2.10; voltage depends not only of the arc length but also on the electrode extension and on the shielding gas.

2.2.3.1 Current

Direct current electrode positive (DCEP) is the most used current in GMAW because it gives stable electric arc, low spatter, good weld bead geometry and the greatest penetration depth.

For low currents and voltages in combination with active shielding gases or mixtures containing active gases, dip or short-circuiting transfer is obtained. Metal is transferred to the work-piece by bridging at frequencies usually above 100 Hz. This metal transfer mode gives low heat input, being suited for welding thin sections and for positional welding.

Globular transfer is obtained for currents and voltages somewhat above those of the dip transfer, if inert shielding gases are used. When carbon dioxide shielding gas is used this metal transfer mode is obtained only for high currents and voltages. Globular transfer is characterized by large drops, with size identical to the electrode diameter or higher, transferred at low frequency. This mode of transfer can be used in a downward direction, due to the predominance of gravitational forces during metal transfer.

The utilization of relatively low current can give insufficient penetration and excessive weld reinforcement, occasioned by poor wetting action of the weld metal. Globular repelled transfer can be found when electrode negative polarity is used with solid wire, but this mode of transfer has no industrial application due to poor stability and high spatter levels which result.

For currents and voltages higher than for globular transfer, projected spray transfer occurs when argon-rich shielding is used. It arises for currents above spray transition current, which depends on the electrode material, shielding gas and electrode diameter. It is approximately 240 A for 1.2 mm diameter carbon steel electrodes with argon/5% CO_2 shielding [3]. This mode of transfer is characterized by very small drops projected onto the work-piece at a very high frequency, up to 350 drops per second, presenting low spatter level. As high currents are used high heat inputs to the work-pieces are reached, producing large weld pools with deep penetration. This type of metal transfer is attractive when high deposition rate welds in thick materials in a downward direction are to be performed. However it presents limited capacity in positional welds, due to the effect of gravity forces. For even higher currents and voltages, streaming spray transfer is obtained, but it has no industrial application due to high weld pool turbulence caused by the increase of the electromagnetic forces.

Drop spray transfer mode can occur in the transition between globular and projected spray transfer, in a restricted operating range. This metal transfer mode is characterized by a very efficient detachment of small drops from the electrode, which are projected onto the work-piece at high velocity and with low spatter level. This type of transfer is difficult to regulate in conventional DC power sources but can be achieved using pulsed transfer techniques.

Pulsed current allows projected spray transfer for mean currents below spray transition current, improving positional capabilities and operating tolerances of the process. Details concerning the control of the metal transfer modes in the arc are given in Chapter 3.

2.2.3.2 Voltage

Arc voltage is directly related to current, as indicated above, and with arc length, increasing with it. Voltage also depends on the shielding gas and electrode extension. The increase of arc voltage widens and flattens the weld bead. Low voltages increase the weld reinforcement and excessively high voltages can cause arc instability, spatter, porosity and even undercut.

2.2.3.3 Welding Speed

Increase in the welding speed gives a decrease in the linear heat input to the work-piece and the filler metal deposition rate per unit of length. The initial increase in welding speed can cause some increase in penetration depth, because the arc acts more directly in the parent material, but further increase in speed decreases penetration and can cause undercut, due to insufficient material to fill the cavity produced by the arc.

2.2.3.4 Electrode Extension

The electrode extension is the electrode length that is out of the contact tube. The increase of electrode extension, produced by the increase of the torch distance to the work-piece for a specific parameters set, increases electrode melting rate because of the Joule effect. Electrode extension ranges from 5 to 15 mm for dip transfer, being higher (up to 25 mm) for the other transfer modes.

2.2.3.5 Shielding Gas

Shielding gases have an effect on arc stability, metal transfer mode, weld bead shape and melting rate. Gases used in GMAW can be pure gases, binary, ternary and exceptionally quaternary mixtures. Common pure gases are argon, helium and carbon dioxide. The first two are inert gases and are used principally in welding of light alloys, nickel, copper and reactive materials. Helium has a higher ionization potential than argon, providing larger weld pools, but is more expensive. Carbon dioxide is an active gas and is used in welding of carbon steels. It produces high levels of spatter but provides high penetration depth.

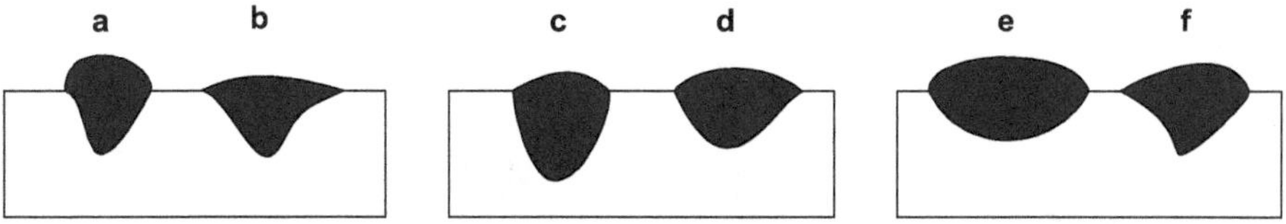

Figure 2.11. Effect of shielding gas on weld geometry. Argon – **a**; argon+oxygen – **b**; CO_2 – **c**; argon+CO_2 – **d**; helium – **e**; argon+helium – **f**

Binary mixtures are commonly argon/carbon dioxide (up to 20% CO_2), argon/oxygen (up to 5% O_2) and argon/helium (up to 75% He). The first is used in the welding of carbon and low alloy steels, the second of stainless steels and the third of nonferrous materials. The addition of oxygen or carbon dioxide to argon stabilizes the welding arc and changes the bead shape [17], as illustrated in Figure 2.11. The objective of adding helium to argon is to increase heat input and consequently welding speed, but also to reduce the incidence of weld porosity.

The most common ternary mixtures are argon/oxygen/carbon dioxide, used in welding of carbon steels, argon/helium/carbon dioxide and argon/carbon dioxide/hydrogen, used in welding stainless steels. Ternary mixtures are intended for improving weld bead profile, increasing tolerance to material contamination and promoting higher travel speeds.

2.2.3.6 Electrode Diameter

Chemical composition of the electrodes is similar to that of the materials being welded. Most usual electrode diameters are 0.8, 1, 1.2 and 1.6 mm. Electrodes of lower diameter are used for thin materials. Electrodes of 1.2 and 1.6 mm diameters are utilized in welding thicker materials and need higher currents, which produce larger weld pools. Electrodes of 1.6 mm diameter are not recommended for positional applications.

2.2.4 Process Variants

Flux cored arc welding (FCAW) is a process similar to GMAW but uses a tubular flux cored electrode as the consumable instead of a solid electrode, as shown in Figure 2.12. Flux has several functions which are deoxidization, alloying, gas generation and formation of a protective slag. The process has two variants, these being the gas-shielding FCAW process, that uses an external shielding gas to assist in shielding the arc and the weld pool from the air, and the self-shielded FCAW process that works without external shielding. Flux-cored electrodes offer several advantages such as higher deposition rate than solid electrodes, because of higher current density of tubular electrodes, alloying addition from the flux, slag shielding and improved arc stabilization, more tolerance to rust and scale than conventional process and the need for less skilled personnel.

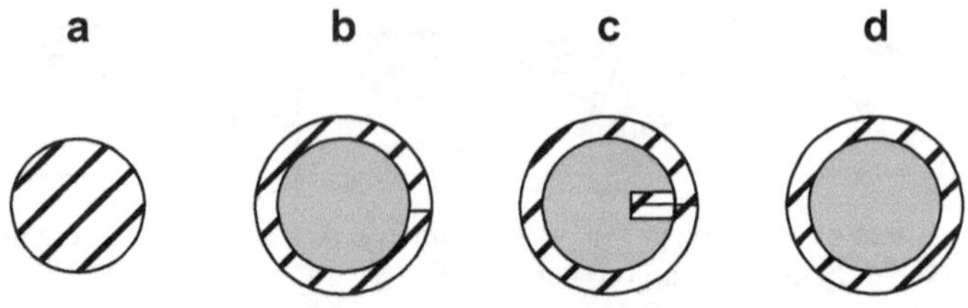

Figure 2.12. Cross section of common flux-cored electrodes. Solid electrode – **a**; flux-cored electrodes – **b**, **c** and **d**

The main limitations of flux-cored electrodes are the large quantity of fumes generated, which is potentially toxic, and the need for removing slag, particularly in multipass welds. Flux–cored electrodes are more expensive than solid electrodes but the difference in cost of the consumable is compensated by the decrease in labor costs because they have higher burn-off rate than solid electrodes.

Flux-cored electrodes of 1 and 1.2 mm diameter can be used in positional work in contrast to electrodes of 1.6, 2.4 and 3.2 mm that must be used in flat and horizontal positions. In the last few years electrodes have been developed mainly for welding carbon and low alloy steels as well as for stainless steels and for hardfacing applications. For steels, CO_2 and argon/CO_2 mixtures are used as shielding gases.

Constant-voltage direct current machines are recommended for FCAW processes, though output rates should be higher than for conventional process. For semi-automatic process outputs, between 400 and 600 A are recommended while for mechanized and robotic systems power sources with outputs, up to 1000 A may be required for some applications. Knurled feed rollers are generally used to feed flux-cored electrodes in order to avoid crushing the electrode, even when using low pressure. Water-cooled torches are used mainly in automatic and robotic welding for currents above 300 A when argon-rich shielding mixtures are used.

MIG/MAG tandem and multi wire welding can give a significant increase in welding speed and disposition rate and also influence the weld geometry [19].

The GMAW process can be used in combination with other welding processes such as plasma arc welding (PAW) or laser welding (LW) to improve deposition rate, welding speed, flexibility and productivity [20],[21]. Limitations of these processes are the high capital cost and complexity in setting optimal welding parameters.

The AC pulsed GMA process is currently under development for robotic welding applications. It is well suited to the welding of aluminum alloys, giving high-quality and productivity in welding of thin-sheet joints. Moreover it extends the root opening tolerance and reduces work-piece distortion, during the welding cycle [58].

2.3 Laser Beam Welding (LBW)

A laser consists of a high-power coherent monochromatic light beam which can be focused to a small spot, producing a very high energy density. Laser is the acronym for "*light amplification by stimulated emission of radiation*". A laser beam is produced by stimulating emission of electromagnetic radiation in specific solid or gaseous materials. Atoms of these materials are moved to higher energy levels by absorbing stimulating energy, producing a population inversion, that is material is brought into a condition in which population of atoms at a higher energy level is greater than that at lower level. These atoms decay by spontaneous emission of photons, which can generate more photons by stimulating emission from other excited atoms, producing the amplification of the laser light. Laser light sources have reflecting mirrors incorporated (see Figure 2.13) which reflect photons back for further light amplification.

2.3.1 Introduction

The most popular lasers for welding are the solid-state lasers of neodymium-doped yttrium aluminum garnet (Nd:YAG), generally pulsed wave, and the gas lasers of continuous-wave carbon dioxide (CO_2), whose lasing medium is a mixture of carbon dioxide, nitrogen and helium. Power density of laser welding (10^9-10^{11} Wm^{-2}) is significantly higher than that of arc welding processes (10^6-10^8 Wm^{-2}), though somewhat lower than electron beam welding (10^{11}-10^{13} Wm^{-2}) [3].

The beam energy delivered to the work-piece will be dissipated by reflection and absorption. Work-piece material is heated to a very high temperature, melted and may even vaporize due to very high power density concentrated in the focus of laser beam. Two modes of laser welding can be obtained, the heat conduction-mode and the deep-penetration mode, depending on the power density in use [22]. Heat conduction-mode is obtained for low power density, where most of the beam energy is lost by reflection (up to 90%), and it is characterized by the formation of a wide and shallow weld pool, see Figure 2.14 a). Power density is sufficient to melt the material but it is not enough to vaporize it, the weld pool shape being controlled by surface tension and thermocapillary forces [23]. This technique is used for welding small components for the electronics industry or for small medical parts.

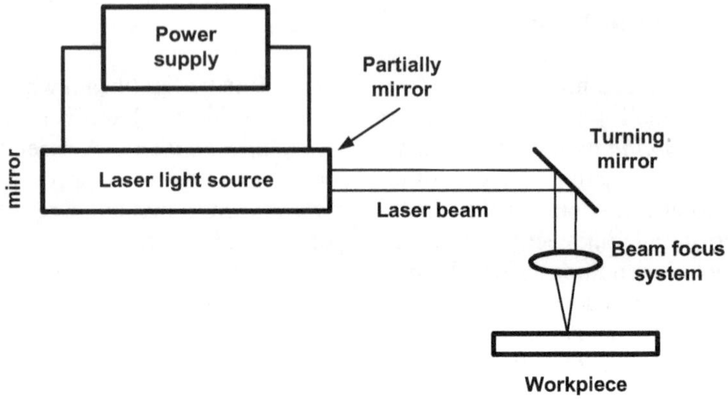

Figure 2.13. Schematic representation of a laser welding system

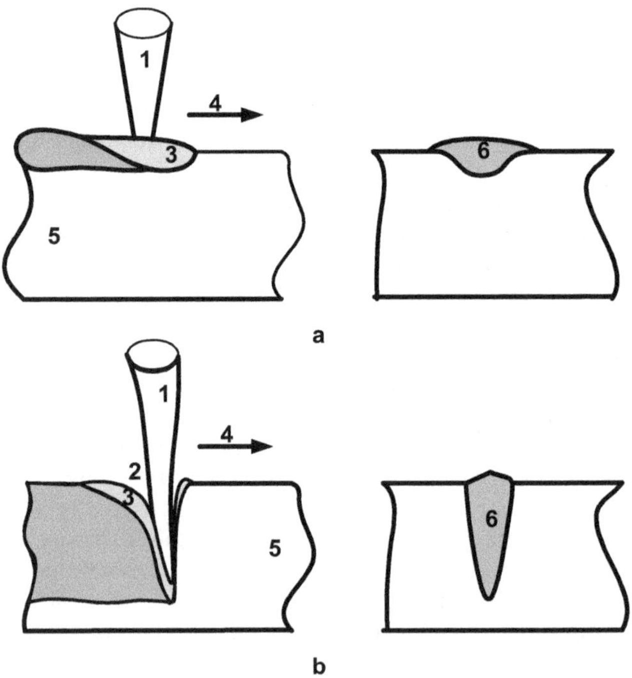

Figure 2.14. Laser welding modes: Heat conduction-mode – **a**; deep-penetration mode – **b**. Laser beam – 1; vapor channel – 2; weld pool – 3; welding direction – 4; work-piece – 5; solid melt – 6.

For power densities above a critical threshold of about 10^{10} Wm^{-2} the laser beam causes melting and vaporization of metal, creating a keyhole in the work-piece, as shown in Figure 2.14 b). The keyhole and plasma generated aid the adsorption of energy by the work-piece and the distribution of heat deep in the material. Metal vapor continuously generated tends to maintain the keyhole while metal flow and surface tension tend to obliterate it. As the laser beam advances it creates a channel and material solidifies behind it. This is the deep-penetration mode laser welding, which produces a narrow and deep welding seam. This welding mode is commonly applied for welding thick materials (up to 50 mm) at high travel speed [22], without filler metal, though filler electrodes can also be used to fill gaps.

The laser welding process provides a high energy density beam that can be used at room atmosphere to produce precise welds at high speed, even in difficult-to-weld materials, such as titanium. Added to this, welds are deep and narrow, with small heat affected zones, giving low distortion, and almost no post processing is necessary [29]. Main limitations of laser welding are the need for accurate part fit-up and precise part positioning as well as equipment capital cost that is ten times more expensive than arc welding systems of identical power. In addition the process is dependent on the material's light absorptivity and surface condition and it is susceptible to weld porosity, solidification cracking and bead geometric defects, mainly in aluminum alloys.

2.3.2 Welding Equipment

The welding equipment includes several types of lasers used in welding. In the following, solid-state lasers and gas lasers will be considered.

2.3.2.1 Solid-state Lasers

Solid-state lasers used in welding are of the ruby type, composed of a ruby crystal containing a concentration of 0.05% chromium, or of Nd:YAG type, made of a solid yttrium aluminum garnet rod doped with neodymium. Excitation of electrons in neodymium is done with high-power xenon flash lamps (1-4 kV), as represented schematically in Figure 2.15. This process is known as pumping. Diode lasers are frequently used as the pumping source instead of flash lamps, in order to improve pumping efficiency. Pumping energy is amplified within the crystal, commonly designated as cavity, which contains a fully reflecting mirror at one end and a partially reflecting mirror at the other. After amplification of radiation the laser beam is radiated from the partially reflecting end, with 1.064 μm wavelength. Because of the limited capacity of cooling systems to maintain a threshold temperature of the crystal Nd:YAG lasers are commercially available up to 6 kW average power, though conventional systems have generally up to 1000 W average power, with a maximum pulse power of 5 to 20 kW, a pulsing rate up to 400 pulses per second and a beam parameter of 25 (mm × mrad) or lower.

Commercial solid state lasers with high pulse power are capable of simultaneous welding at several different locations. The weld point diameter can also be adjusted by the processing optics at a constant working distance of 0.1 to 2 mm, and the welding depth can be controlled via the laser parameters up to 2 mm.

The Laser beam can also be transmitted through fiber optics which leads to several advantages, such as improved flexibility of laser systems and reduced need for accurate mirror alignment.

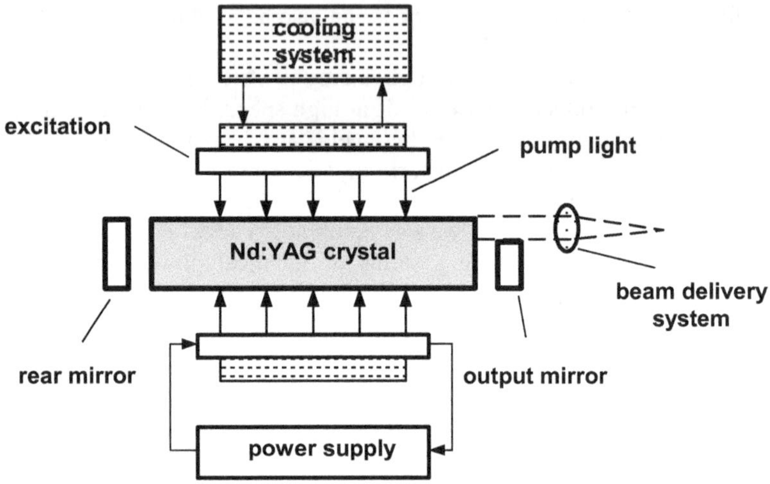

Figure 2.15. Schematic representation of a Nd:YAG laser system

2.3.2.2 Gas Lasers

Gas lasers have several characteristics different from solid lasers. The radiation wavelength of CO_2 lasers is 10.6 μm and the transmission of the laser beam is made by reflection using mirrors. They can be used in pulsed or continuous modes, in a power range up to 25 kW, though lower powers are more usual.

Axial flow CO_2 lasers are composed basically of a laser tube where the gas mixture flows, the front and rear mirrors and the radio frequency electrodes for excitation of the laser gas. The rear mirror is fully reflecting, opposite to the front mirror where a partially reflecting window exists. Windows of germanium or gallium arsenide are used in order to transmit laser beam without significant loss. The most usual laser gas mixtures are composed of carbon dioxide (5%), nitrogen (15%) and helium (80%) or oxygen (3.5%), carbon dioxide (4%), nitrogen (31.5%) and helium (61%). The gas mixture must be water cooled, because an increase in gas mixture temperature can cause decomposition of carbon dioxide and a decrease in efficiency of the laser. These lasers are called slow axial-flow lasers and are limited to small powers (500 W). In modern laser systems the heat generated in the gas is dissipated by the water-cooled electrodes (diffusion-cooled). A beam shaping module is integrated into the laser head and produces a high quality round symmetrical beam. The resonator design produces a 45° linearly polarized beam [29]. Output power up to 4.5 kW can be obtained with these lasers.

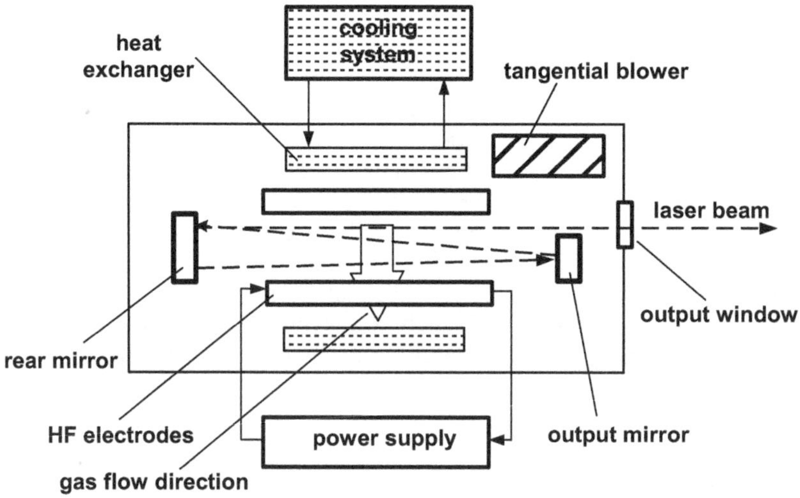

Figure 2.16. Schematic representation of a CO_2 transverse-flow laser system

In fast axial-flow lasers gas in the laser tube is re-circulated at high speed by blowers or turbines and heat removed by a heat exchanger. These lasers are composed of several optical units in series, in order to increase output power, with the optical resonator being folded several times to obtain a more compact system. The laser beam is transmitted between optical units by intermediate mirrors. Output powers up to 5 kW can be obtained with this type of laser.

In transverse-flow lasers gas is circulated into the discharge region transversely across the line of discharge by a tangential blower, being cooled by a heat exchanger, see Figure 2.16. This arrangement results in compact lasers, allowing shorter resonant cavities and higher outputs than axial-flow lasers. Power outputs up to 8 kW can be obtained with these lasers. Most of these lasers can be used with either continuous wave or pulsed wave, with variable pulse frequency between 0 and 100 kHz.

2.3.3 Process Parameters

Primary parameters of laser welding are the beam power, the beam diameter and travel speed, though other aspects, such as the control of plasma formation, the welding gases and the absorptivity of the parent material, can have drastic effect on weld penetration depth and on metallurgical changes in the weld.

2.3.3.1 Beam Power and Beam Diameter

Penetration depth increases almost linearly with increase of power density, for a specific diameter of the laser beam. Power density depends on the power of the laser beam and on the focus cross section area. Beam diameter is very small and it is difficult to evaluate because energy in the beam normally has a Gaussian distribution. This distribution is designated as the transverse electromagnetic $mode_{00}$ or TEM_{00}. Conventional definition of the beam diameter is based on the diameter where power density is $1/e^2$ of maximum power in central part. The circle defined in this way contains 86.5% of the total beam energy [3]. Other beam energy distributions may be observed, such as doughnut distributions, but they are not beneficial for welding operations because of the decrease of coherence of the beam.

2.3.3.2 Focus Characterization

Focus is basically characterized by the minimum focal spot size (d_{min}) and the focus depth (Z). Focal spot size is relevant to the determination of power density and its theoretical value can be determined by Equation 2.2, where f is the focal length of the focusing optics, λ is the wave length of the laser beam and D is the diameter of the unfocused beam, as illustrated in Figure 2.17. Frequently focused beam diameter is larger due to imperfections of the focusing optics [24].

$$d_{\min} = \frac{1.27 f\lambda}{D} \qquad 2.2$$

Focus depth is defined, according to *Laser Institute of America*, as the distance in which focus spot radius is increased by 5%. Focus depth can be estimated by Equation 2.3, where F equals $f\lambda$ of the optic system:

$$Z = 1.488 F^2 \lambda \qquad 2.3$$

Focus depth increases with increase of the F number of the focusing optics but focus diameter also increases, decreasing power density. Focus depth is important when welding thin components because thermal distortion can put beam focus out of these components.

The position of focus has great influence on quality of welds produced. If focus is well above the surface of the work-piece, welds show a nail head appearance and little penetration is obtained. When focus is positioned deep below the work-piece surface V-shaped welds result and a more accurate setting of the components is needed. Optimum focus positioning is below the work-piece surface but distance is a function of plate thickness and beam power.

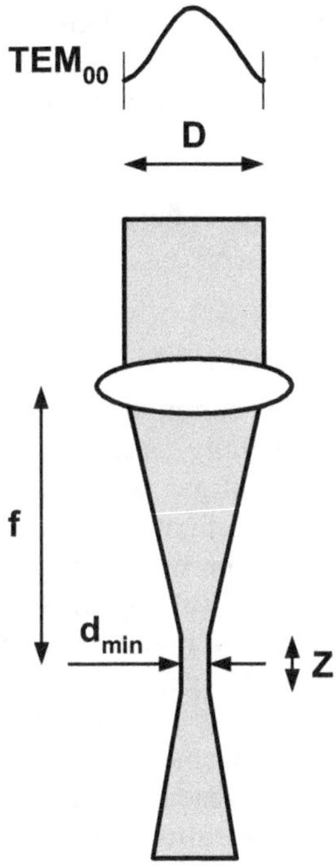

Figure 2.17. Characteristic parameters of focal system

2.3.3.3 Travel Speed

The increase of travel speed decreases penetration depth for both argon and helium shielding gases. This is because power input per unit length decreases with welding speed increase and keyhole may not be completely effective in trapping incident radiation. Very high speeds can give lack of fusion while low speeds may originate excessive parent material melt, vaporization and even defects formation. For very low speeds a reduction in penetration may be observed. This is attributed to the formation of a cloud of plasma, which attenuates the incident laser beam.

2.3.3.4 Plasma Formation

For power densities above 10^{10} Wm^{-2} in CO_2 lasers and 10^{12} Wm^{-2} in YAG lasers, the beam interacts with metal vapor and shielding gas, producing a cloud of plasma above the plate surface. During the initial moments of keyhole formation, plasma may assist the energy transfer to the work-piece. However, subsequently, plasma may limit beam energy transfer to the work-piece [25]. Several techniques have been developed to reduce plasma formation or to remove it from the weld zone. Pulsing laser power at high frequencies (above 1 kHz) is effective in reducing plasma formation in CO_2 lasers. In addition plasma is generally removed from the vicinity of the beam by an auxiliary jet of helium or argon. Assisting gas must be directed to 1 mm ahead of the beam, at an angle of approximately 20 degrees with the work-piece surface. Helium is preferred, because it has a higher ionization potential than argon, being more resistant to plasma formation. Beam interaction with the work-piece can also be improved by the combination of linear oscillation of the beam in welding direction with the jet of an inert gas [3].

2.3.3.5 Welding Gases

In laser welding two gases are commonly needed, the assisting gas to remove plasma, which is injected laterally, and a coaxial shielding gas to prevent atmospheric contamination. A root gas is also needed in keyhole welds where all the material thickness is melted. Helium and mixtures of argon and helium are used as welding gases. Argon shields the weld metal and helium is required to control the plasma formation in CO_2 laser welding. If Nd:YAGs are used for welding, the plasma formation is not an aspect of major concern and argon is the recommended welding gas. Small additions of oxygen, hydrogen or CO_2 can be used depending on material and process to increase productivity further [26]. Helium, argon or mixtures of these gases are used for most materials, including reactive metals such as titanium or zirconium. For reactive materials the shielded area must be increased, because they are sensitive to air contamination down to low temperatures (400 °C). Nitrogen can also be used for welding stainless steels in less demanding applications [3].

2.3.3.6 Absorptivity

The efficiency of laser beam welding represents the proportion of beam energy that is effectively added to the work-piece. It is drastically affected by the absorptivity of the material to be welded. Absorptivity is a function of the electrical resistivity of the material, according to Equation 2.4, where A is the absorptivity and p_r the the electrical resistivity

$$A = 112.2\sqrt{p_r} \qquad 2.4$$

Absorptivity in many metallic materials is very low, 2 to 3% for aluminum or copper and less than 15% for stainless steel [22]. Absorptivity is increased by the formation of oxide layers in metallic materials. Absorbent powders can be applied in work-piece surface, in order to reduce reflection losses. The addition of active gases, such as oxygen, to shielding gas also improves absorptivity. In keyhole welding absorptivity suffers a large increase because of multiple reflections inside the keyhole, providing efficient welding even in high reflective materials such as aluminum [27].

The beam energy absorbed by a specific material is also a function of the radiation wavelength, generally increasing with the decrease of the wavelength. For steels absorptivity of Nd:YAG radiation is approximately three times of that of CO_2 laser radiation. For aluminum this difference is not so large and for other materials, such as copper or silver, no difference exists in this range of wavelength.

2.3.4 Process Variants

Dual beam laser welding has been proposed few years ago to improve fit-up tolerances and to reduce the probability of forming bead shape defects, such as humping and undercutting [59]. Beams can be mounted side-by-side or the second beam trails behind the primary beam.

Robotic hybrid welding processes were also developed to increase welding speed and deposition rate. This is the case for the combination of laser and GMAW processes. This combination provides high speed and good fit-up tolerance.

High power lasers, such as CO_2 lasers, needed for high speed welding of metals, require large floor space, considerable electrical and water services and regular maintenance. For precision welding applications, a new generation of lasers named diode lasers is available, providing a more efficient operation and maintenance-free running for more than 10000 h [28]. In fact it is not a variant but a new type of laser. These lasers incorporate diode chips, each one emitting a laser beam of very low power, when excited electrically. These chips are mounted into bars containing a cooling system and micro-channel lenses to focus individual laser beams. These bars have low power, around 60 W, and are mounted into diode stacks with other optical systems in order to obtain a focused laser beam with a power of several kW [29], as represented schematically in Figure 2.18. These lasers can be classified as low power diode lasers (LPDL), having power up to 150 W, and high power diode lasers (HPDL) with power ranging from 150 W to 4 kW. The lasers of this last group are used in welding operations. The wavelength of the laser beam is in the range 0.63 to 0.99 μm, though the interval 0.8 to 0.94 μm is common in welding applications. Aluminum has a marked increase of absorptivity in this wavelength range. Diode laser beam is not as coherent as Nd:YAG or CO_2 laser beams and focus is larger and rectangular instead of circular, being less sensitive for fitting of components to weld.

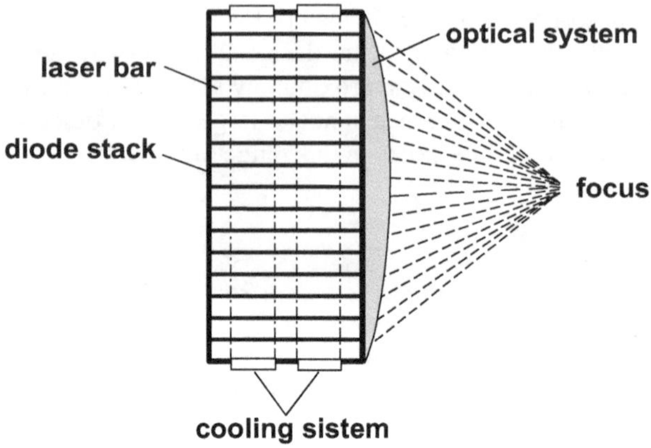

Figure 2.18. Schematic representation of a diode laser

This process has high energy efficiency (30-50%) when compared with CO_2 lasers (3-10%). Added to this, HPDL are compact and light and can be easily adapted to anthropomorphic robots with small pay-load (less than 25 kg). Running cost are approximately one-tenth of CO_2 lasers but beam quality is low [31]. HPDL are applied to the welding and brazing at high speed of carbon and stainless steels and aluminum alloys, as well as cladding operations. Thickness of welded components is limited by the power of the laser. They are becoming increasingly used in welding of thermoplastic materials, where they are replacing traditional techniques such as ultrasonic welding [32].

2.4 Resistance Spot Welding (RSW)

Resistance spot welding (RSW) is included in the group of resistance welding processes in which the heat is generated by passage of electric current through the bodies to be joined, according to Joule's law, expressed by Equation 2.5, where H is the heat generated, I is the current and t is the time of current flow:

$$H = I^2 R t \qquad 2.5$$

Other welding processes such as resistance seam welding, projection welding, flash or upset welding and high-frequency welding are of the same group. Spot welding is the resistance welding process most widely used in robotic applications all over the world and is treated here with some detail. Main aspects of resistance seam welding process, which also has some relevance in industrial robotics, are analyzed in the section of process variants.

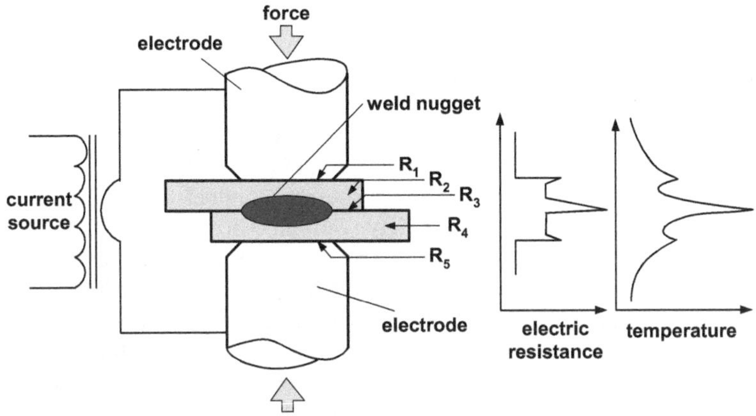

Figure 2.19. Schematic representation of the spot welding process. Electrode-work- piece interface resistances – R_1 and R_5; resistance of the work-pieces – R_2 and R_4; resistance in the interface between work-pieces – R_3

2.4.1 Introduction

In resistance spot welding overlapping sheets of metal are joined by applying electric current and pressure in the zone to weld with copper electrodes, as illustrated in Figure 2.19. Copper is used for electrodes because it has low electrical resistance and high thermal conductivity. Spot welding operation is composed of three steps that are the squeezing, welding and holding stages. Squeezing consists of applying the weld force to the work-pieces in order to obtain the appropriate amount of pressure, prior to welding. During welding, the electric current passes through the work-pieces, while the welding force is maintained, generating heat. In the course of the holding stage current is switched off and weld force maintained, allowing the weld to forge and cool under pressure.

The heat generated depends basically on the electrical current and time being used and on the electrical resistance of materials between electrodes. This inter-electrodes resistance is composed by five separated resistances, as is indicated in Figure 2.19. Resistances R_1 and R_5 are undesirable because they produce heating and consequently degradation of the electrodes. Resistances R_2 and R_4 are the resistances of the work-pieces and they assume particular importance in the final period of the weld. Low resistive materials are difficult to weld because of reduced heat generated in the pieces. Resistance R_3 is the most important because it determines nugget formation, assuring the establishment of the weld.

The nugget is a volume of melted material that forms in the interface of work-pieces with a diameter similar to that of the electrodes, as is indicated in Figure 2.19. Nugget penetration should be al least 20% of the thinnest sheet member but not exceeding 80% of the same thickness [31]. The passage of current initiates after the application of the electrodes force, leading the increase of temperature in the interface and developing a molten nugget. In the final part of the welding cycle plastic deformation occurs in the work-pieces, producing a visible and permanent indentation of the pieces. If current or pressure is too high, melted material can be expulsed (splashed) to the atmosphere.

The process has extensive application in welding of carbon steels because they have higher electrical resistivity and lower thermal conductivity than the electrodes made of copper. Aluminum alloys have an electrical resistivity and thermal conductivity that are closer to those of the copper, making difficult the welding operation of these materials, requiring higher levels of current, which can damage the electrode tips [32]. Other materials such as galvanized steels, heat-resisting alloys and reactive metals are also welded by this process. Since the process is very competitive it is widely used in automotive and aerospace industries as well in the manufacture of industrial and domestic equipment.

The major advantages of this process are the high welding speed and low thermal distortion, respectively faster and lower than in conventional arc welding processes, suitability for automation, the need of low skilled operators and the absence of joint preparation or filler metal. Some limitations of this process are the need for lap joints in thin materials, usually up to a thickness of 4 mm, the joints are not tight and have low tensile and fatigue strengths. Add to this fact that the initial equipment costs are higher than those of conventional arc welding equipment.

2.4.2 Welding Equipment

The main welding equipment to consider in resistance spot welding are the welding power sources and the electrodes. Those pieces of equipment will be considered next in detail.

2.4.2.1 Power Sources

Spot welding machines are composed basically an electrical circuit, which provides welding current, a control circuit that regulates welding current and welding time, and a mechanical system, used to apply welding force.

The electrical circuit consists of a step-down transformer, whose secondary circuit includes the electrodes and the work-pieces, see Figure 2.19. The transformer changes the input AC high-voltage and low amperage current, in the primary winding, to an AC high-amperage and low voltage current in the secondary winding. These transformers have low internal impedance, because current magnitude in secondary winding is inversely proportional to the impedance and depends directly on the open voltage of the secondary circuit.

Single-phase AC machines providing current up to 50 kA are widely used. Spot welders can also provide DC of continuous polarity, pulses of current of alternating polarity or pulsed mode [15]. Single or three-phase machines are available, though single-phase are commonly used because they are simpler to operate and have lower initial and maintenance costs for almost equivalent performance. Three types of direct current machines are generally available: the rectifier, the frequency converter and stored energy machines. The rectifier and frequency converter machines are fed from three-phase systems in contrast to stored energy machines that draw power from single-phase systems. These latter machines store energy during a period of time and then discharge a pulse of current to make the weld. These welders are useful for low frequency welds. Medium-frequency (400-2000) DC inverters are available for RSW. These inverters improve ability to control the welding process finely [60]. High-frequency DC inverters are being developed for further improvement of the process control.

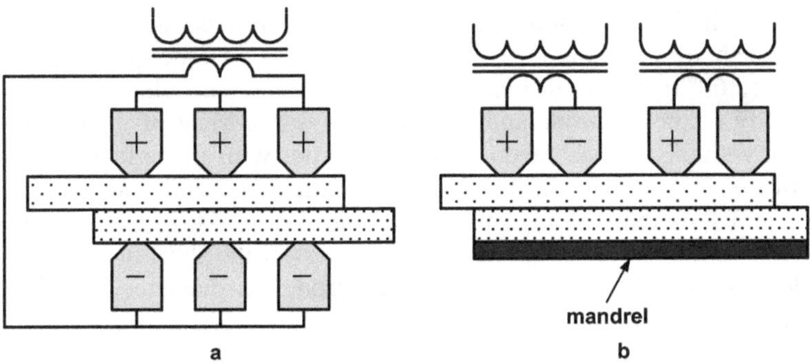

Figure 2.20. Arrangements of the secondary circuit for multiple spot welds; **a** - direct welding; **b** - series welding

In multiple spot welding the arrangement of the secondary circuit depends on whether they are direct or series welds, as is illustrated in Figure 2.20. In direct multiple-spot the welding conditions are similar in the three electrodes represented in Figure 2.20a; in series multiple-spot each of the two transformer secondary circuits shown in Figure 2.20 b makes two welds [33].

Most of the spot welders are computer controlled and allow the input of welding data. The simplest control sets current magnitude and welding time. More sophisticated controls allow to regulate current during welding as well provide and control preheat and post-heat operations [34].

Electrode clamping force is applied by hydraulic, pneumatic, magnetic or mechanical means, at a high controlled velocity in order prevent premature deformation of the electrodes. During the welding cycle, material clamped by the electrodes expands and contracts rapidly, because of the high heating and cooling rates, but working pressure must be maintained. When heated metal undergoes softening the electrodes must follow-up to maintain enough pressure on the sheet surfaces. If pressure drops during welding electrode-work-piece interface resistances increase, electrodes are overheated and may deteriorate. Clamping force can be variable during the cycle. Metals which have high shrinkage during solidification may need an increase of force to forge nugget after current passage. Modern systems allow control of the clamping force during all stages of the spot welding process. New portable gun units, incorporating the power transformer and the actuator into a common platform, facilitate the fitting of RSW to robotic systems.

2.4.2.2 Electrodes

Electrodes should have high electrical and thermal conductivities and must develop low electric contact resistance in order to prevent deterioration of the work-piece and electrodes. In addition they must have good strength to resist to deformation and wear at high temperature. They are made from copper containing alloy elements such as chromium (0.6-0.9%), cobalt (1-2%), beryllium (0.5%) and zirconium (0.08%). Electrodes are composed of three parts: the electrode cap or tip, the body of the electrode and the cooling system. Most of the electrodes are cylindrical with the tip machined to a truncated cone with an angle of 30º, though a variety of tip shapes (pointed, flat, dome and radius) is used to obtain access with complex joints. When welding thin sheets the electrode tip diameter (*d*) can be estimated by Equation 2.6, where t is the thickness of the sheet in contact with the electrode. For thicknesses above 1 mm electrode tip diameter is estimated by Equation 2.7:

$$d = \sqrt{t} \qquad \text{(a)} \qquad 2.6$$

$$d = 5\sqrt{t} \qquad \text{(b)} \qquad 2.7$$

When welding sheets of dissimilar thicknesses the electrode diameter should be specified for the thinner material. The weld diameter (*D*) should be similar to the electrode diameter. Repeated heating and cooling of the electrode and pick-up of metal particles causes electrode tip deterioration. Maximum tip diameter allowed is 1.3 times the initial diameter. When the electrode tip reaches this diameter it should be replaced or redressed to the original diameter.

Electrodes should be water cooled at a flow rate above 4 l/min and separate water circuits must be used for both top and bottom electrodes.

2.4.3 Process Parameters

According to Joule's law the welding parameters are current, time and electric resistance. Electric resistance is a function of several parameters such as electric resistivity of materials, surface quality of sheet metal and clamping force. Electric resistivity is a characteristic of the materials and varies with temperature during welding. Material surface quality depends on roughness and cleanliness of the surfaces of sheet materials. Parameters that can be programmed in the welding machine are current, time and welding force. The choice of welding conditions depends on thickness and physical properties of metals being welded and even on the type of the welding equipment.

2.4.3.1 Welding Current and Time

Heat developed during welding is proportional to time and to square of current. Though both parameters are responsible for heat generation, the weld heating rate is determined only by current, because heat lost to the work-piece and to copper electrodes increases with weld time. Heat lost to the work-piece increases heat affected zone and thermal distortion, while heat in the electrodes can degradate them, all being undesirable effects. The level of current required for any metal tends to be inversely proportional to its electrical and thermal resistivities.

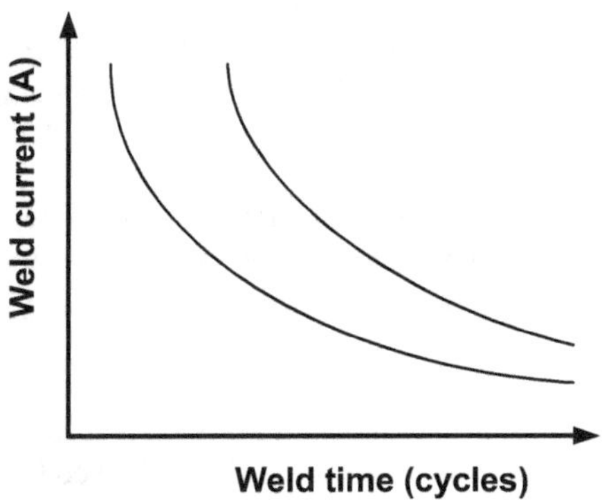

Figure 2.21. Schematic representation of current-time relationship for RSW

The size of the weld nugget increases rapidly with increasing current. When welding a particular material and thickness, if current is increased welding time should be decreased, see Figure 2.21, in order to prevent high surface indentation or even expulsion of melted material and deterioration of electrodes [35]. The expulsion of material defines the upper limit of usable current.

Welding currents range from 20 to 100 kA, mainly for light alloys, though the most usual are between 4 and 20 kA for carbon steels. Time is defined in cycles of 50 Hz supply and it is between 5 and 100 Hz for steels and 5 and 20 Hz for light alloys, in sheets up to 3 mm thick.

Weld current cycles may have different shapes that depend on the materials being welded, as shown in Figure 2.22. A cycle of constant current magnitude, Figure 2.22a, represents the simplest situation and is suitable for welding mild steels. For high strength steels sensitive to cold cracking a modulated welding current with a rise time t_r and a fall time t_f, see Figure 2.22b, can be used to allow gradual heating and cooling of the weld. In the case of materials prone to form brittle structures in the weld an additional current cycle of magnitude I_a, see Figure 2.22c, can be useful to anneal the weld. In spot welding of thick materials (over 3 mm) the use of several pulses of current, Figure 2.22d, is effective.

2.4.3.2 Welding Force

The increase of the welding force reduces contact resistance because, in first analysis, it promotes the increase of contact area, due to deformation of surface asperities and eventually the rupture of surface oxide films [36]. Electrode clamping forces must be high, particularly when welding low resistivity metals in order to reduce the proportion of heat generated in the interface electrode/work-piece. Electrode force must be increased with increasing current, unless part of the melted material of the nugget can be expelled. Other factors such as bad fit and lack of mechanical support contributes for the material expulsion. Distance of the weld to the edge of the sheets should be larger than 1.5 D, where D is the weld diameter. Excessively high forces are also undesirable because they can cause large surface indentation of the work-pieces and damage of the electrodes.

Electrode clamping force increases with increasing thickness and strength of the work-pieces. Forces between 1000 and 15,000 N are usual for plate thicknesses up to 3 mm, though values of 20,000 N can be used in steel sheets 6 mm thick.

Clamping force starts before the passage of current initiates and is maintained after the current is cut off, as is illustrated in Figure 2.22. Sometimes an increase of force is applied after current passage to forge the weld, see Figure 2.22b.

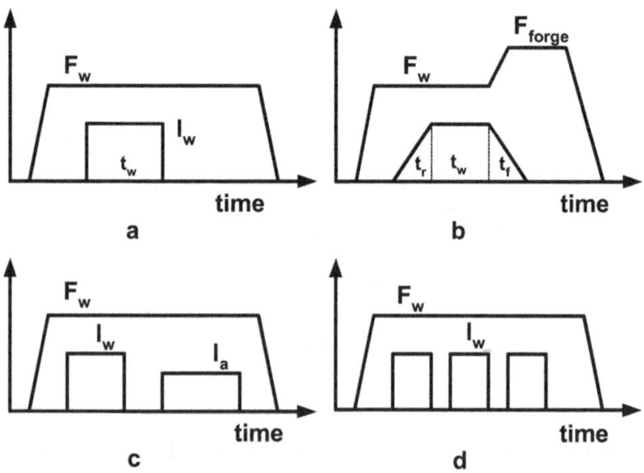

Figure 2.22. Timing diagrams of current and force for spot welding: Welding current – I_w; welding time – t_w; rise time – t_r; fall time – t_f; welding force – F_w; forge force – F_{forge}; annealing current

2.4.4 Process Variants

Resistance seam welding (RSEW) is used when a continuous seam is required. This seam consists of a series of overlap spot welds, as shown in Figure 2.23. This process is similar to resistance spot welding, but the electrodes are replaced by power driven wheels or rollers that move along the joint. Electric current passes intermittently while the wheels are stationary, without the necessity of raising or lowering the welding head. The amount of overlap between spots is 25-50%. The process can be used to do spot welds by simple adjustment of timing. The weld width in continuous welds is between $2\sqrt{t}$ and $5\sqrt{t}$, where t is the single sheet thickness. The track tends to deform due to continuous work and a device is needed to correct the shape of the wheel edge.

RSEW machines can be of circular type, where the axis of rotation of the electrode wheel is at right angles to the front of the machine, of longitudinal type, where the the axis of rotation of the electrode wheel is parallel to the front of the machine, and of universal type, which allows the orientation of the axis of rotation of the electrode wheels to be changed [32]. Portable machines are also available for welding large work-pieces that are difficult to handle by conventional equipment.

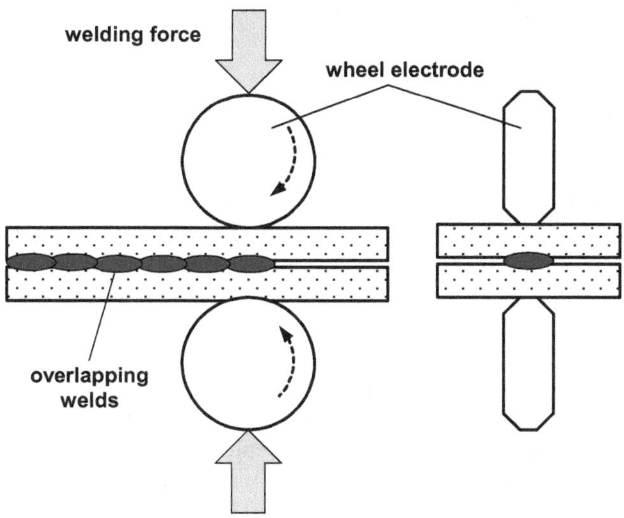

Figure 2.23. Seam welding principle

Electrode wheels are made of the same materials of RSW electrodes, with diameters between 50 and 610 mm and can have internal or external cooling. Internal cooling may have higher operational costs and do not cool the weld.

Maximum welding current in conventional RSEW machines ranges commonly from 20 to 30 kA, though welders up to 100 kA are applied in welding of light alloys. Clamping forces between 2000 and 16000 N and welding speeds ranging from 1 to 12 m/min are used for steels, though lower values of force and speed are applied in aluminum alloys.

Carbon, low-alloy, stainless and coated steels are currently welded using this process. Welding of light alloys requires additional precautions because of their lower electrical resistivity and lower melting temperature. A new process named conductive heat resistance seam welding allows one to increase the welding speed and reduce joint preparation cost in difficult-to-weld aluminum alloys [37].

RSEW is largely used in the automotive industry as well as in manufacturing of heat exchangers, non-pressurized tanks and several types of cans.

Main advantages of this process when compared with resistance spot welding are the capacity to produce gas-tight and liquid-tight welds as well as the possibility of reduction of the overlap width of the sheets. However, the weld must progress in a straight line or in a uniformly curved line of large radius and thermal distortion can be higher than in resistance spot welding.

This process has several variants such as mash-seam welding, butt seam welding, high frequency resistance welding and high frequency induction welding [38] but they are outside the scope of this introduction.

2.5 Friction Stir Welding (FSW)

Friction stir welding (FSW) is a solid state joining process invented at *The Welding Institute (TWI)* in 1991 [39], in which a non-consumable rotating tool is slowly plunged into the butting faces of the work-pieces and traversed along the joint line, see Figure 2.24a. Pieces to be welded have to be clamped in order to prevent joint faces from being moved out of position.

Heat is generated by tool friction, under the tool shoulder and on the probe surface, and by plastic deformation of the material [40]. Heat generated is lost to the work-pieces, to the tool and to the anvil, as represented schematically in Figure 2.24b. Maximum temperatures are attained close the tool shoulder and are lower than the melting temperature of the materials being welded [41], though incipient melting has been reported for some materials. Heat produced creates a softened plasticized region around the tool, which facilitates the movement of the tool along the joint line. Plasticized material is chaotic mixed or extruded from the advancing side to the retreating side of the tool [42],[43] and it is forged by the contact of the tool shoulder and of the pin, producing a solid phase bond between the two pieces.

2.5.1 Introduction

This welding process leads to the appearance of a thermo-mechanically affected zone (TMAZ), which results from both plastic deformation and thermal exposure of the material, and of a heat affected zone (HAZ), which only suffers the effect of the thermal cycle. In the central part of the TMAZ there usually appears a distinct nugget, having an onion ring feature, attributed to dynamic re-crystallization or dynamic recovery of the microstructure [44].

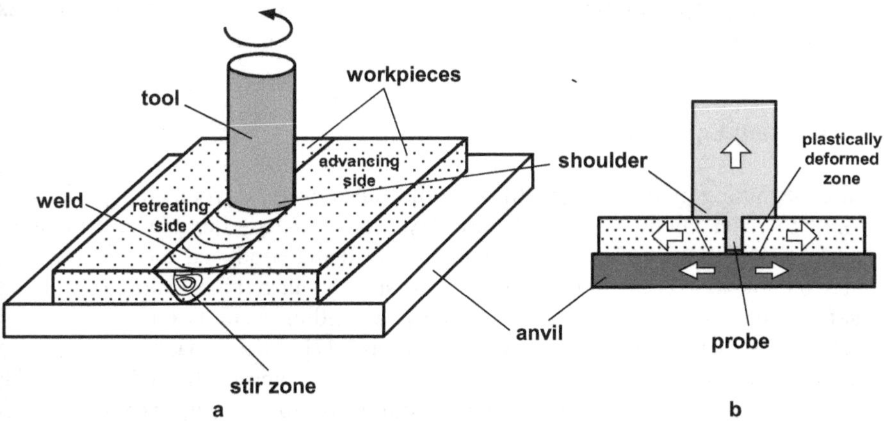

Figure 2.24. Schematic representation of the friction stir welding process

FSW is mainly used in welding of aluminum alloys, though other materials such as magnesium, copper, zinc, titanium and even steel [44] can be welded with this process. This process can be used too for welding aluminum alloys of different alloy groups or yet dissimilar materials, metal matrix composites and plastics. It presents several advantages when compared with conventional arc welding processes, mainly in the welding of aluminum alloys. Difficulties related to sensitivity to solidification cracking, gas porosity caused by the hydrogen absorbed during welding and thermal distortion, very common in fusion welding processes, do not happen in this process. Other benefits of the process include good strength and ductility along with minimization of residual stress and distortion. These qualities of FSW are generally attributed to the solid-state nature of the process and a supposed low energy input to the welded joint. In addition to this no filler electrode, no shielding gas and minimum surface preparation is needed. No environmental concerns have to be considered because neither fumes nor toxic gases nor radiation of the electric arc are produced in this process.

However, there are still several drawbacks that need to be addressed in order to facilitate industrial application of this process. The system requires high forces to move the tool through the plasticized material, which in turn wears the tool, mainly in welding of hard materials. Powerful clamping fixtures are also needed to hold pieces down and counteract forging forces from the tool. Because of this, FSW is usually carried out in custom heavy-duty machine tool equipment, where weld joints are frequently limited to straight lines or two-dimensional contours. The use of industrial robots increases the flexibility of this process, providing the ability to weld three-dimensional contours [46].

Nowadays FSW is used in the welding of pieces in aluminum alloys ranging in thickness from 0.5 to 75 mm. It is being used in the shipbuilding and the marine industries, for manufacturing of panels, platforms and heavy profiles, in the aerospace industry for production of fuel tanks, wings and fuselages, in the railway industry for high speed trains, in the automotive industry, for production of panels and other components, *etc.*

2.5.2 Welding Equipment

In the beginning mainly high stiffness machines were developed, specifically tailored to meet client needs, but nowadays standardized, flexible and modular systems suited to several industry segments are being produced too, for welding nonferrous metals. These latter systems consist of a sturdy basic framework, a set of safety stops, a welding carriage assembly, a welding head assembly, a control system, a hydraulic unit and the welding tools [47]. These machines can have several basic designs, providing vertical down forces ranging from 6 to 200 kN, welding speeds up to 2 m/min, though an option up to 6 m/min exists, and tool rotation speed between 500 and 2000 rpm.

As referred above these machines have low flexibility producing welds in simple two or three-dimensional pathways. Robotic systems allow the improving of flexibility but need to be able to apply and maintain a large and constant axial force during the welding operation, which is not simple in these systems. This is done using high payload robots that sense the force directly and use feedback to maintain the force during the welding operation. The axial force decreases with increasing tool rotation speed but increases with increasing travel speed, and therefore for robotic FSW a compromise may need to be established between travel speed and axial force requirements [46].

The appropriate tool type is a key factor of the quality of friction stir welded joints. For butt welding aluminum alloys of thickness up to 12 mm cylindrical threaded pin probes are recommended, while for thicker plates the Whorl and MX-Triflute probes should be used [48], see Figure 2.25. These latter probe types allow welding speeds that exceed largely those achievable with threaded pin probes: at least by a factor of 2. In addition they have flat or re-entrant features or oval cross section, which reduce the probe volume (static volume), allowing one to achieve a suitable swept volume (dynamic volume) to static volume ratio. The greater this ratio, the greater the path for material flow and the efficiency of the probe [49].

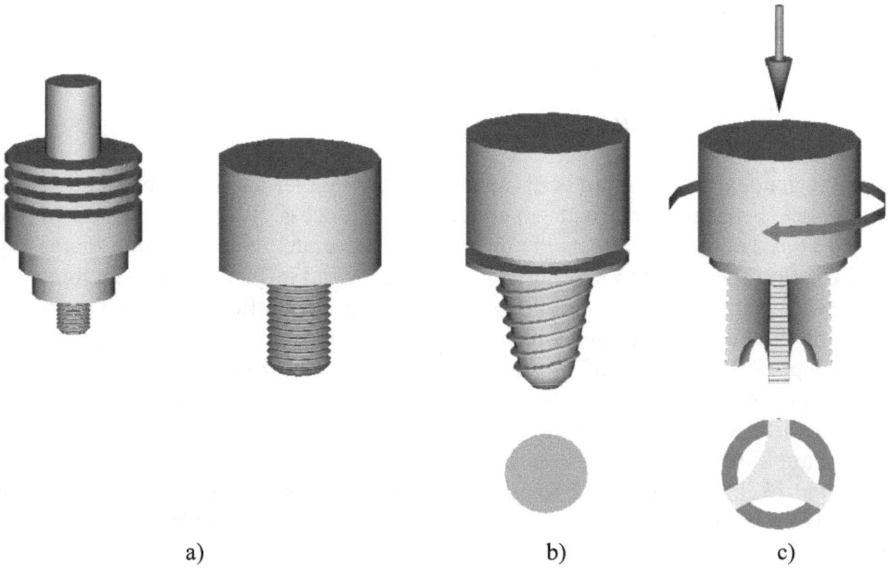

Figure 2.25. Friction stir welding probes. Cylindrical threaded pin probe – **a**; oval shape Whorl probe - **b**; flared-triflute probe – **c**

For lap welding of aluminum alloys, which is more difficult than butt welding because wider welds are necessary and oxide disruption at the sheet interface is more difficult, special tools are used, such as Flared-Triflute and A-Skew probes [49]. These tools allow the dynamic to static volume ratio to be increased, improving weld quality.

The shape of the bottom of the tool shoulder affects material flow around the probe and contributes to preventing the escape of plasticized material. They can be flat or concave, smooth or grooved, with concentric or spiral grooves [50]. A concave shoulder bottom has the advantage, when compared with a flat bottom, of directing material flow to the center close the probe. Grooved bottoms have in general the same effect. Shoulders can also have knives to shave the weld.

In general this welding process requires access to both sides of the work-pieces being welded, although by using a special tool (bobin-tool) it is possible to do the welds without the need for an anvil [51].

2.5.3 Process Parameters

The main parameters of the FSW process, which are determinant for the quality of the welded joint, are the vertical down force, also designated tool plunge force, the tool rotation speed and the travel speed or welding speed. Plunge force assures penetration of the probe into the plates and forges plasticized material under the shoulder. The tool rotation speed directly influences the heat generated in the process because the mechanical power input to the tool is given by Equation 2.8, where P is the power, M the torque and Ω is the angular speed of the tool.

$$P = M\Omega \qquad 2.8$$

The mechanical power input is dissipated mainly by thermal losses because plastic work may be neglected. Heat generated in the process is also influenced by the plunge force because it affects the torque. Heat generated increases with increasing tool rotation speed and tool plunge force. Travel speed influences the heat input per unit weld length (specific heat input), affecting metal flow around the probe. Specific heat input decreases with increasing travel speed, which reduces material softening in the vicinity of the probe, making plastic flow more difficult. High travel speeds may cause defects, such as cavities. For low tool rotation speed, low plunge force and high travel speed external defects may form for welds in some aluminum alloys. The increase of the plunge force moves defects to the interior of the weld [52]. The ratio tool rotation speed *vs* travel speed is sometimes used to distinguish between hot welds, having high ratio, and cold welds, with low ratio. Hot welds are less sensitive to defect formation but may exhibit more significant changes in microstructure and mechanical properties than cold welds in aluminum alloys.

Other relevant parameters are the time of indentation of the tool and the tool shoulder angle, besides the other geometric characteristics of the tool referred in the previous section. The time of indentation of the tool is the period between the instant the tool contacts the work-piece and the instant the tool begins moving along the joint. During this period generated heat spreads in the vicinity of the probe, softening material and stabilizing material flow around the probe. If this period is too short defects can appear in the initial part of the weld. Time can range usually from 5 to 30 s. The tool shoulder angle allows a gradual increase of the pressure on the top surface of the plates being welded and helps to direct the material flow. Tool angles up to 3° are common.

2.5.4 Process Variants

In the last few years several variants of FSW process have been developed. One of these variants is thermal assisted FSW in which a heat source is applied in the joint before the FSW tool, in order to preheat and soften the material [53]. This reduces welding forces, welding power and tool wear and increases travel speed. This variant can be useful in welding of steels and other high strength materials.

Another variant is spot FSW, developed for lap joints, that produces spot welds having higher mechanical strength than those produced by resistance spot welding. Robotic applications of this process are being developed [54].

A recent development is the reversal stir welding process (Re-Stir™), developed by TWI, in which tool rotation is applied as both angular reciprocating, where reversal is imposed within one revolution, and rotary reversal, where reversal is imposed after one or more revolutions, instead of continuous rotation as is in conventional FSW. Re-Stir is basically a cyclic and essentially symmetrical process. According to TWI Re-stir™ may become the preferred option for certain butt, lap, compound lap and spot welding and material processing applications [55].

2.6 Health and Safety

The major potential hazards of arc welding processes are the high-voltage electricity, which can injure and kill personnel, the fumes and gases, which can be dangerous to health, the electric arc radiation, which can injure eyes and burn skin and the noise that can damage hearing.

The exposure to the high open-circuit voltage of power supplies can cause dangerous electric shocks, which can be prevented by connecting all the electrical equipment and work-pieces to a suitable electrical ground. All electric cables should be suited to the maximum current and must remain insulated and dry.

Fumes and gases are generated in all arc welding processes, being particularly intense in the flux cored arc welding process. Metal fumes of nickel, chromium, zinc, lead or cadmium, for example, and gases such as carbon monoxide, ozone and nitrogen oxides formed in the arc are very harmful to the health [56]. Enough ventilation or exhaust at the arc, or both must be used in order to keep fumes and gases from the personnel breathing zone.

The electric arc of GTAW and GMAW processes emits intense radiation in the ultraviolet range, in the infrared range and also in the visible range. UV radiation can commonly cause a temporary eye burn, which can be painful for 48 h. A filter glass should be used by the operator to absorb the radiation in the dangerous wavelengths, and limit visible light so he can see the joint during the welding operation. There are two basic types of filter, permanent filters and photosensitive filters, which react rapidly to the incident light from the arc and darken [55]. Optical density of filters increases with increasing current. The UV also occasions reddening and irritation of the skin and operators need to be protected by leather, wool or aluminum coated clothing. Robotic welding systems are generally protected by enclosures provided with windows with filters for viewing weld area.

Ear protection should be used when noise is excessive in the work area. Special care must be taken in handling and use of cylinders containing high-pressure and liquefied gases, which should remain in a vertical position, secured with chains, when they are being used.

Lubricants or other flammable compounds should not be used in pressure-reducing regulators and other parts of the oxygen circuit because they can lead to catastrophic fire.

Potential hazards in laser beam welding are in many aspects similar to those observed in arc welding. Laser power supplies employ high voltage capable of producing lethal electric shocks. Laser welding also generates dangerous metal fumes, whose composition depends on the metals being welded, requiring local exhaust ventilation. However laser beams can cause permanent eye damage, so exposure to direct or reflected laser beams must be prevented. Laser welding systems must operate in restricted access enclosures opaque to the laser wavelength. Individual laser eye protection can be required for personnel working in the vicinity of the laser source [22]. Thermal burns can also occur if skin is exposed to primary laser beams.

Principal motives of concern in resistance spot welding are protection against molten metal spatter and splash and electric shock. Working environment can be improved by the use of enclosures and splash-less resistance spot welding systems [57].

2.7 References

[1] Aoki A,.Takeichi M,.Seto M,.Yamaguchi S, (2004), Development of GTAW robot system for aluminum frame, IIW Doc XII-1814-04.

[2] Johnson M,.Fountain C,.Castner H, (2000), GTAW fluxes for increased penetration in nickel based alloys and titanium, IIW Doc. XII-1617-00.

[3] Norrish, J, Advanced welding processes, Institute of Physics Publishing, 1992

[4] Hichen G K, Gas-tungsten arc welding, ASM Handbook, Vol 6, Welding, Brazing and Soldering, pp 190-194.

[5] Chen Y, Nie ZR, Zhou ML, Zhang JX, Zuo TY. The research and development of tungsten electrodes without radioactivity, International Symposium on Ecomaterials held in conjunction with the 39th Annual Conference on Metallurgists of CIM, AUG 20-23, 2000 ENVIRONMENT CONSCIOUS MATERIALS - ECOMATERIALS, 699-702, 2000

[6] Tusek J., Suban M. (1999), TIG welding in a mixture of argon, helium and hydrogen, IIW SG-212-948-99.

[7] Shirali, A. A., Mills, K. C., The effect of welding parameters on penetration in GTA welds, Welding J. 72(7) 1993, pp. 347s-353s.

[8] Lancaster, J.F., Mills, K.C., Recommendations for the avoidance of variable penetration in gas tungsten arc welding, IIW Doc 212-796-91.

[9] Pierce, S. W., Burgardt, P., Olson, D. L., Thermocapillary and arc phenomena in stainless steel welding, Welding J., 78(2) 1999, pp. 45s-52s.

[10] Paillard, P., Saindrenan, J., Effect of activating fluxes on the penetration capability of the TIG welding arc: study of fluid-flow phenomena in weld pools and the energy concentration in the anode spot of a TIG arc plasma, Materials Science Forum 426-4, Ed. Chandra, T., Torralba, J.M., Sakai, T., 4087-4092, 2003.

[11] Watanabe H., Butsusaki Y., Nagashima T (2004), Development of ultra-narrow gap hot wire GTA welding process, IIW Doc. XII-1810-04.
[12] GTAW process with dual shielding gas, IIW Doc. XI-455-86, 1986.
[13] Jarvis, B.L., Ahmed, N.U., Sc. and Tech. Weld. Joining, 2000, 5(1), 1-7.
[14] Holliday, D B, Gas-metal arc welding, ASM Handbook, Vol 6, Welding, Brazing and Soldering, pp 180-185.
[15] Grist, F J, Farrel, W and Lawrence, G S, Power sources, ASM Handbook, Vol 6, Welding, Brazing and Soldering, pp. 36-44.
[16] Hancock, R and Johnsen, M, Developments in guns and torches, Welding J (2004), 83(5), pp. 29-32.
[17] Lyttle, K A, Shielding gases, ASM Handbook, Vol 6, Welding, Brazing and Soldering, pp.64-69.
[18] Temrat P, Poopat, B., Preliminary study of effect of GMAW's pulse shape on weld profile, Proceedings of the IIW Asian Pacific International Congress, Singapore, 29 Oct. - 1 Nov. 2002, publ. by The Welding Technology Institute of Australia (WTIA), ISBN 0-909539-99-5, vol. ST 3/4, paper N° 37.
[19] Bohme D., Nentwig A., Knoch R., A high efficiency welding process - the double wire welding, Proceedings of the 1996 IIW International Congress, Auckland, 07-09 February 1996, pp. 1393 – 1407
[20] Rayes, M, Walz, C and Sepold, G, The influence of various hybryd welding parameters on bead geometry, Welding J (2004), 83(5), pp. 147s-153s.
[21] Messler, R, Whats's next for hybrid welding, Welding J (2004), 83(3), pp. 30-34.
[22] Mazumder, J, Laser beam welding, ASM Handbook, Vol 6, Welding, Brazing and Soldering, pp. 263-269.
[23] Zacharia, T, David, S.A., Vitek, J. M. and Debroy, T., Weld pool development during GTA and Laser beam welding of type 304 stainless steel, Part I and II, Welding J. 68(12) 1989, pp. 499s-519s.
[24] Mazumder, J, Procedure development and practice considerationsfor laser-beam welding, ASM Handbook, Vol 6, Welding, Brazing and Soldering, pp. 874-880.
[25] Katayama S., Seto N., Mizutani M., Matsunawa A. (2000), Interaction between plasma and laser beam, and its effect on keyhole dynamics in high power C02 laser welding, IIW Docs IV-765-00, SG-212-976-00
[26] Faerber, M.G., Gases for laser cutting and welding, Proceedings of the 2000 IIW International Congress, Melbourne, 29 Oct. - 2 Nov. 2000
[27] Kawahito, Y., Katayama, S., In-process monitoring and adaptive control in laser micro-spot lap welding of aluminum alloy, Proceedings of the IIW 2004 Int. Conf. "Technical trends and future prospectives of welding technology for transportation, land, sea, air and space", Osaka, 15-16 July 2004, pp. 433-438.
[28] Kraft, T., Chang, J., Hoult, A., Lee, S., Migliore, L., New advances in laser materials processing, Proceedings of the 2000 IIW International Congress, Melbourne, 29 Oct. - 2 Nov. 2000 (ICRA-2000-37).
[29] ROFIN Laser Diodes, General catalog, http://www.rofin.com (Laser Diodes), 2004.
[30] Bryden, B. G., Welding of plastics with high power diode laser, Industrial Robot 31(1) 2004, 30-33.
[31] Linnert, G. E., Welding Metallurgy, Fourth Edition (1994), AWS.
[32] Matsuyama, K., Evaluation of electrode tip life using various types of aluminum alloy sheets for automobile body in resistance spot welding with dome and radius electrode tips, IIW Doc III-1125-98.
[33] Procedure development and process considerations for resistance welding, ASM Handbook, Vol 6, Welding, Brazing and Soldering, pp.833-850.
[34] Welding Handbook, Volume 2, Welding Processes, Eigth edition, AWS.

[35] Seo, D. W., Jeon, Y.B., Lim, J.K., Effect of electric weld current on spatter reduction in spot welding process, 5th Int. Conf. on Fracture and Strength of Solids/2nd Int. Conf. on Physics and Chemistry of Fracture and Failure Prevention, Oct. 20-22, 2003, Advances in Fracture and Failure Prevention, PTS 1 and 2, 1623-1628, 2004
[36] Chang, B.H., Zhou, Y., Numerical study on the effect of electrode force in small-scale resistance spot welding, 9th Int. Man. Conf., Aug 16-17, 2000, J Mater Process Tech 139 (1-3) 2003, 635-641.
[37] Kimchi, M., Workman, D., Gould, J.E., Advanced welding techniques for aluminum alloys, Welding in the World, July 2002, vol. 46, Special issue, pp. 157-168.
[38] Karagoulis, M.J., Resistance seam welding, ASM Handbook, Vol 6, Welding, Brazing and Soldering, pp.238-246.
[39] Thomas, W. M., Nicholas, E. D., Needham, M. G., Templesmith, Dawes, C. J., Friction stir butt welding, International patent application PCT/GB92/02203, GB Patent application 9125978.8. US Patent 5.460.317, 1001.
[40] C. M. Chen, R. Kovacevic, Finite element modeling of friction stir welding – thermal and thermomechanical analysis, Int. J. Machine Tools and Manufacture 43 (2003) 1319-1326.
[41] Khandkar, M. Z. H., Khan, J. A., Reynolds, A. P., Prediction of temperature distribution and thermal history during friction stir welding: an input torque based model, to appear in Science and Technology of Welding and Joining.
[42] Colligan, K., Material flow behaviour during friction stir welding of aluminum, Welding J. 78(7) 1999, pp. 229-237.
[43] Li, Y., Murr, L. E., McClure, J. C., Flow visualization and residual microstructures associated with the friction-stir welding of 2024 aluminum to 6061 aluminum, Materials Science and Engineering A271 (1999), pp. 213-223.
[44] Fonda, R. W., Bingert, J. F., Colligan, K. J., Development of grain structure during friction stir welding, Scripta Materialia 51 (2004), pp. 243-248.
[45] Thomas, W. M., Threadgill, P. L. and Nicholas, E. D., Feasibility of friction stir welding steel, Science and Technology of Welding and Joining (1999), vol. 4 (6), pp. 365-372.
[46] Cook, G. E., Smartt, H. B., Mitchell, J. E., Strauss, A. M., Crawford, R., Controlling robotic friction stir welding, Welding J. 82(6) 2003, pp. 28-34.
[47] ESAB Genaral Catalog, www.esab.com, 2005.
[48] Thomas, W.M. and Dolby, R. E. 2003, Friction stir welding developments, Proc. 6th Int. Conf. on Trends in Welding Research. Eds. S. A. David, T. DebRoy, J. C. Lippold, H. B. Smartt and J. M. Vitek, pp. 203-211. ASM International.
[49] Thomas, W. M., Staines, D. G., Norris, J. M., Frias, R., Friction stir welding – Tools and development, FSW Seminar, Porto, Portugal, TWI 3/12/02.
[50] Dawes, C., Thomas, W, Development of improved tool designs for friction stir welding of aluminum, 1st Int. Sym. On Frictio Stir Welding, Thousand Oaks, California, USA (1999).
[51] Strombeck, A., Schilling, C., dos Santos, J., Robotic friction stir welding, GKSS Workshop, Reibruhrschweissen, Geesthacht, Germany, 2002.
[52] R. Leal and A. Loureiro; Defects formation in friction stir welding of aluminum alloys; Materials Science Fórum Vols. 455-456 (2004) 299-302.
[53] Kohn, G., Greenberg, Y., Makover, I., Muntz, A., Laser-assisted friction stir welding, Welding J. 81(2) 2002, pp. 46-48.
[54] Sakano, R., Murakami, K., Yamashita, K., Hyoe, T., Fujimoto, M., Inuzuka, M., Nagao, M., Kashiki, H., Development of spot FSW robot systems for automobile body members, 3rd Int. Sym. on Friction Stir Welding, Port Islands, Kobe, Japan 2001.
[55] The Welding Institute, UK, http://www.twi.co.uk/
[56] http://www.cpwr.com/hazpdfs/kfwelding.PDF

[57] Matsuyama, K. I., Improving working environment in resistance spot welding, IIW Doc III-1248-03.
[58] Ueyama, T., Tong, H., Harada, S., Passmore, R., Ushio, M., “AC pulsed GMAW improves sheet metal joining”, Welding J. 84(2) 2005, pp. 40-46.
[59] Xie, J., “Dual beam laser welding”, Welding J. 81(10) 2002, pp. 223s-230s.
[60] Villafuerte, J., “Advances in robotic welding technology”, Welding J. 84(1) 2005, pp. 28-33.

3

Sensors for Welding Robots

3.1 Introduction

The demand for better control and sensing in welding has increased with automation and welding processes involving new and advanced materials. This requires precise control of the welding process to produce the desired weld with respect to productivity and quality. Consequently, there is a need for different technologies to control precisely the process with respect to the different welding operating parameters. In doing so, sensors play a crucial role as the major source of input to the control system that manage and control the behavior and output of the welding system. Here, the term "*sensor*" is used for devices that measure observable parameters related to the welding process which is used to control the process in accordance with defined specifications.

In general, most robotized welding processes that produce a continuous weld are based on the MIG/MAG process, or GMAW (Gas Metal Arc Welding). Within this application field, the use of sensors has been modest. However, the development and introduction of new welding processes like high speed welding, laser welding, *etc.*, emphasize the importance of accurate control of the process. The development of new products also makes use of new materials with possibilities to decrease thicknesses. A result of this is a need to be able to work with tighter tolerances. Thus, the need is increasing for sensors that can meet the requirements from new processes and product specifications and is in many cases a necessity.

The main task of the sensors is to provide the control system with information to generate proper actions to produce a result that corresponds with defined specifications. In welding this is not as easy as we might think. From a welding process perspective, the process is performed mainly by two subsystems; the welding equipment and the robot. The welding equipment includes the welding power source and the devices that deliver the energy from the welding power source, like the wire feed system, conduit, welding torch and so on. The robot produces the relative positioning of the energy and the work-piece that is to be welded through a weld torch attached to the end effector mounting plate.

From a control point of view, both the welding equipment and the robot are important to produce the weld with the specified quality and productivity. They are normally controlled by two different and loosely coupled control systems, controlling the welding power source and the robot arm. In the following, both the sensors and the purpose of utilizing sensors to control the weld process will be discussed. Concerning sensors, they are in most cases used for one of the control systems, the welding equipment or the robot, but for the purpose of using sensors, the information can preferably be used for both controlling the welding power source and the robot arm as will be discussed later in this chapter. However, the purpose of the sensor and how it will be used will affect the specification of the sensor which therefore can be divided into two groups, technological and geometrical sensors, measuring technological parameters with respect to the welding process and geometrical parameters with respect to the weld joint geometry.

Sensors that measure geometrical parameters are mainly used to provide the robot with seam tracking capability and/or search capability, allowing the path of the robot to be adapted according to geometrical deviations from the nominal path. Technological sensors measure parameters within the welding process for its stability and are mostly used for monitoring and/or controlling purposes.

As will be discussed later, information from both technological and geometrical sensors provides a basis for a qualitative control of the welding process to make it possible to conform to specifications defined within a WPS (Welding Procedure Specification) concerning quality and productivity measures. Another issue of importance is the mapping problem between observable parameters and controllable parameters with respect to the sensor. In the simple case with a sensor based seam tracking of almost straight welds, the feedback control is straightforward, but applying the sensor data for integrated control related to the WPS will require a more sophisticated model-based control approach. This is because a controllable parameter is not always a parameter that a sensor can observe during the welding process and thus, a model-based mapping must be made to be able to control the weld. In this context, a model-based approach may transform a parameter or set of parameters not directly observable by the sensor into the known set of parameters which can be calculated from a model of the welding process using the information of the sensor. However, the welding process includes many interrelated parameters with included tolerances, which means that such models will predict a number of data not observed directly and with a degree of uncertainty. In practice, such models work better close to defined nominal data where process conditions are similar to those anticipated in the WPS. In the same way, actions to control the process are usually defined by a combination of a set of parameters which together counteract deviations from specifications defined in the WPS. In general, such a set of parameters are not unique and there are in most cases several possibilities to control the welding process and fulfill requirements defined in the WPS.

It should be noted here that the concept of specifying the weld to be produced within a WPS document is a method to define a procedure to, in a consistent and reliable way, work out the welding process. It should include information about the actual weld to produce, like joint geometry and material, but also joint preparation, consumables like shielding gas composition and flow, and welding wire, and nominal operating parameters and the productivity and quality to achieve. In most cases, however, we can only measure some of the parameters needed and from the available observations make judgments on how to control the process to reach the specifications defined in the WPS.

In this way, the WPS is both a specification of the functional specifications of the weld (quality, productivity) and operating data (nominal operating parameters) to reach the specifications. To use sensors, however, means to measure the real process, extract features from these measurements and, through a control action, override these pre-set parameters in order to achieve the functional specifications, where quality and productivity issues have to balance each other on a case by case basis. For a typical case, nominal operating data is defined based on specified weld quality and productivity. This is used to pre-set data of the welding power source and to instruct the robot to generate the trajectory in accordance with defined velocity, welding torch orientation with respect to the weld joint and the distance to the weld joint. If a sensor is used which through a feed-back loop will alter one or several of these settings, the WPS should include allowed tolerances for all nominal data. This is also the case if sensors are used for the purpose of monitoring the robotic welding operation.

3.1 Sensors for Technological Parameters

Technological parameters include voltage, current and wire feed speed. In this section sensors to measure those parameters are reviewed.

3.1.1 Arc Voltage

The measurement of the arc voltage should, in principle, be made as close to the welding arc as possible. The current is delivered to the wire at the contact tube and this could be assumed to be a good measuring point for arc voltage. However, there is a voltage drop between the contact tube and the wire tip where the arc starts of about 0.3 V, depending on the process characteristics [1]. In practice, it is difficult, if not impossible in a production environment, to measure the true arc voltage. This is also the case for measuring the voltage at the contact tube in the weld torch, and a better and more reliable way is to measure the voltage on the wire inside the wire feeding system. As the wire does not carry a current between the feeder and the contact tube the voltage at the wire feeder will be the same as that at the contact tube. It must be noted that measurements are made within an environment that uses high currents, usually in the range of 150-500 A, and if wires used for sensors are placed in the wrong way, this can result in substantial induced voltages with corresponding reading errors.

3.1.2 Welding Current

Basically there are two types of sensors for measuring the welding current: *Hall Effect* and *Current Shunt.*

3.1.2.1 Hall Effect Sensor

The *Hall Effect* sensor consists of a circular core of cast iron through which the cable that carries the current flows. The device itself is placed in the gap in the iron core which in turn consists of a doped silicon plate with two pairs of connecting cables. The first pair feeds the device with a current and the device then responds by delivering a signal on a second pair of cables which is proportional to the magnetic field and thereby the current. The benefit of the *Hall Effect* device is that it is a non-contact sensor and does not interfere with the current of the welding power source. The sensor is limited in bandwidth which is usually in the order of 100 kHz or more and a typical slew rate is 50 A/μs This is usually sufficient and will, provided the data read I/O channels are designed in accordance with the specification of the sensor, result in a rise time of less than 1% of the peak pulse time during pulsed GMA welding.

3.1.2.2 Current Shunt

The principle is to let the current flow through a resistor and measure the voltage drop across it, as if measuring a current with a multimeter. The method is simple but has some drawbacks. The resistor must be kept low and hence the voltage signal measured will be small and sensitive to noise.

3.1.3 Wire Feed Speed

The wire feed speed is a major parameter to control to achieve a stable welding process. The welding power source is in most cases controlled to produce a constant voltage and the preset parameters are usually voltage and current. However, in reality, a current will represent a certain wire feed speed and the common method is to apply a constant value of voltage and wire feed speed and let the current adjust itself accordingly. A higher wire feed speed will produce a higher arc current that will deposit the wire faster and *vice versa* with a smaller value of the wire feed speed. Modern welding power sources have various control schemes, but the general principle is to keep the voltage at a preset constant value. As a result, the current will vary during the process as a result of changes in the environment of the arc welding process, like changes in the distance from the weld torch to the work-piece.

These changes can occur as a result of the movement of the robot or geometrical variations in the weld joint. However, changes of speed can also originate from the wire feeder system and in the end produce poor results with respect to quality.

The wire is fed through a conduit and the wire feed speed is generated by a feeding unit. The unit usually has driving wheels that feed the wire through the conduit at a specified speed. In normal robot systems, the wire feed unit is mounted on the robot arm rather close (the order of 1 m) to the weld torch, giving a reliable push-feed of the wire during the welding process. However, in some cases, longer conduits must be used and since the wire diameter must be slightly less than the inner diameter of the conduit, a varying wire speed will result during welding when the conduit is bent and twisted. In practice, a push-pull wire feed system should be used to counteract this problem.

Measuring the wire feed speed is a big problem and for laboratory purposes custom-made solutions can be build that measure the speed at the contact tube. A more realistic approach in a production system is to measure the controlled speed of the drive wheel of the feeder unit. However, this must be complemented with securing the functionality and reliability of the feeder system as its robustness is important for the resulting quality of the welds.

3.2 Sensors for Geometrical Parameters

Sensors for geometrical parameters must be able to obtain information about the weld that relates to the geometry of the weld joint. This information is of great importance in order to both perform seam tracking and use this information for quality control of the weld. This can be done in many ways but in most cases, a seam tracker is able to extract some information about the weld joint besides the target positions for the weld joint during welding. Examples of such data include deviations from a nominal path, orientation changes and gap size.

To apply sensors in robotics welding means, in general, to use the sensors during welding. In some cases sensors can be used to measure the position and orientation of the weld joint or work-piece before welding. In such cases, different location techniques can be applied in a similar way as for locating any work-piece such as image recognition or binary sensors to detect the position of some plates of the work-piece.

The challenge, however, is to use sensors during welding. Due to the harsh environment with high temperature, intensive light and high currents, purpose built sensors must be applied. The most common use of sensors are (i) optical sensors that use a laser light source on the weld joint under study and a sensor with a narrow bandwidth filter to extract the information of interest, and (ii) through-arc sensing that uses the electrical parameters from the arc together with knowledge about the motion of the weld torch which is controlled by the robot.

3.2.1 Optical Sensors

Optical sensors use the following basic principle for detecting the weld joint during arc welding; (i) a laser beam that is projected in a scanning motion across the seam and (ii) a CCD-array that is used to measure features of the weld joint in combination with a laser stripe. Variations of this method are in use and, as an example, the laser stripe may not be a linear line on the weld joint but circular instead. In such a case, the sensor is more flexible to detect weld joints in corners from one location of the torch, or point of view of the sensor. To measure the distance, the method of triangulation is used which is of great importance in welding, see Figure 3.1.

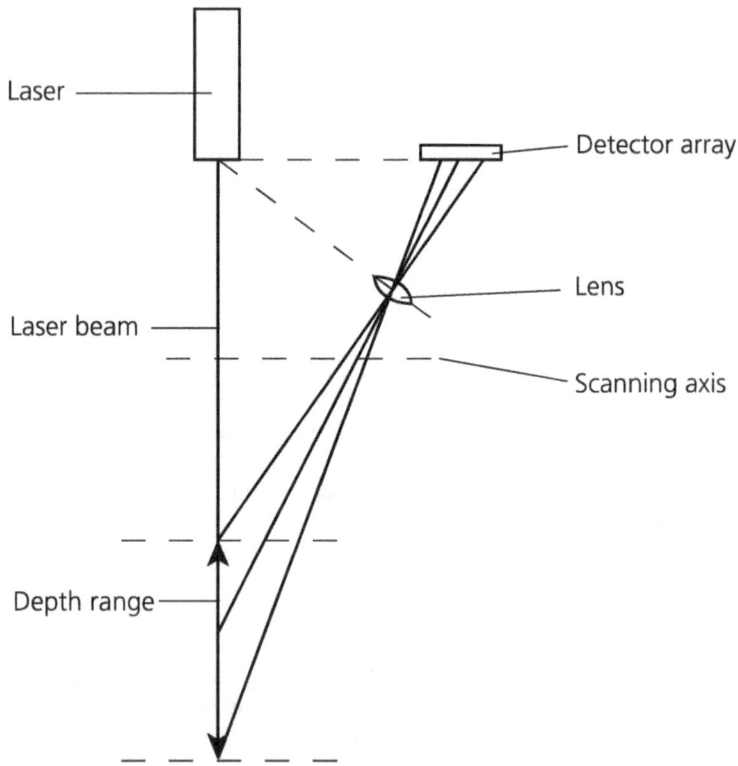

Figure 3.1. The working method of the triangulation method [2]

A laser beam is focused on an object, and then the reflection from the object as seen from a lens in the laser sensor is determined by the distance between the sensor and the object. If the object is close to the sensor then the angle between the outgoing beam and the reflection through the focusing lens of the detector is large, while it is small if the object is farther away. The detection of the distance between the sensor and object is made by focusing the incoming beam on a detector, in most cases a CCD array. Depending on which of the pixels of the array are illuminated, it is possible to calculate the distance to the object.

Depending on the weld joint preparation and geometrical shape, the laser beam can produce reflections like mirrors. Consider for example a V-groove weld joint where the laser light will produce several reflecting positions but with different intensities depending on the surfaces of the weld joint. Therefore, these sensors must have real time image processing capabilities to filter out reflections that do not belong to the point of interest. It should be noted in this context that highly reflective materials may cause problems during welding and a real test may be needed to verify the functionality.

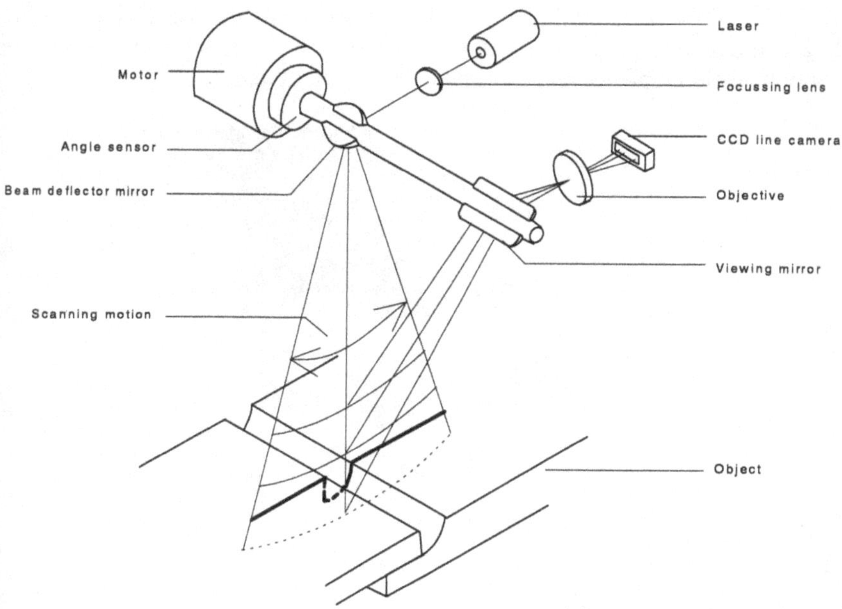

Figure 3.2. Scanning principle of a seam tracking combined with the triangulation method [2]

The basic functionality of a triangulation sensor is to measure the distance to the spot of the object the beam is pointing at. In some cases this can be useful, *e.g.* to control the height during an operation of a robotized process like welding or cutting. But the general use of triangulation in welding is for seam tracking and this requires measuring the weld joint geometry. This is achieved through a scanning technique of the beam across the weld joint, see Figure 3.2. During the scanning, the sensor acquires a two-dimensional picture of the joint profile as an array of 2D coordinates. When the robot is moving, a weld joint geometrical model can be made that contains a full 3D description of the joint which is created during the welding operation when the sensor is moved along the joint.

If a laser stripe is projected onto the object and sensed by a 2D CCD array, the image information can be used directly without moving the robot. This is a technique that is useful if the stripe is circular and aimed at a corner. Then, the corner and its walls can be located from one position of the robot only, compared with the more time consuming traditional technique of measuring the location of one wall at a time.

In most cases, optical seam trackers based on triangulation are used to keep the robot "on track" with the weld joint during welding in real time. However, these sensors have a capability for more than that, and information that can in most cases be achieved that include joint volume, gap size, misalignment, tack welds, *etc.* This information is useful for adaptive feed-back control of both the welding power source and the robot to perform the task in accordance with the predefined specifications of the welds to be produced. As an example, the travel speed of the welding gun defined by the robot can be controlled with respect to the gap of the weld and the welding power related parameters in combination.

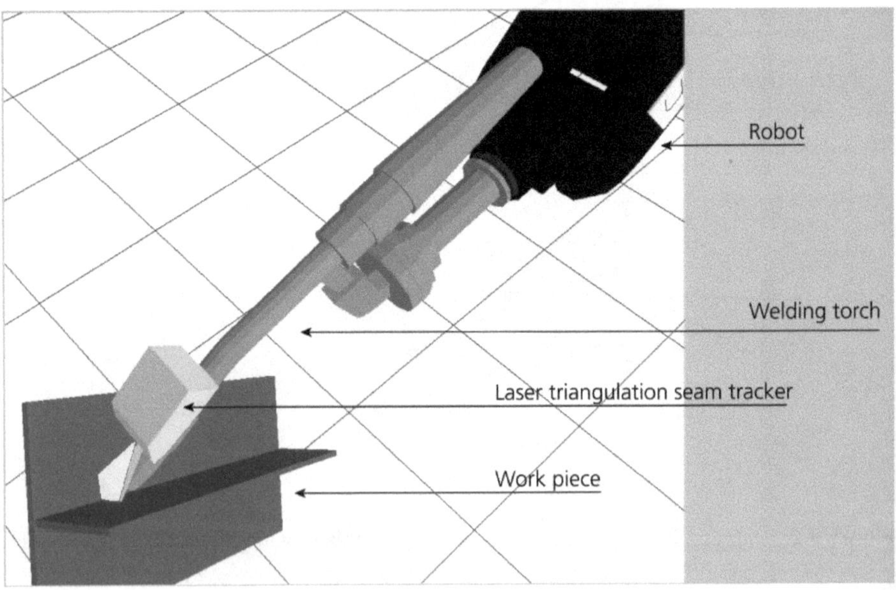

Figure 3.3. Illustration of a typical laser scanner sensor [3] mounted ahead of the welding torch

If the gap varies, there are a few things to control: (i) the metal deposition should be done in a way to obtain a constant weld shape, (ii) a large gap is sensitive for burn-through and a control action could be to lower the travel speed together with lower wire feed speed (weld current), (iii) a too low current may however result in a lack of fusion and cracking in the weld and to keep a higher current a weaving motion of the welding torch can be applied to avoid the problem with burn-through. This shows that the control scheme needs to consider many issues, of which some are boundary conditions and some are counteracting each other, leading to new ways to perform the weld operation.

In robotic welding of thicker plates, the welding is usually performed in several passes, *e.g.* one root pass and additional passes to fill up the weld joint. In such cases, the use of a tracking sensor can be applied in several ways. Normally, tracking is applied for the root pass. During this operation the robot records the weld path and subsequent passes can be overlaid with respect to the first path based on the actual weld joint geometry.

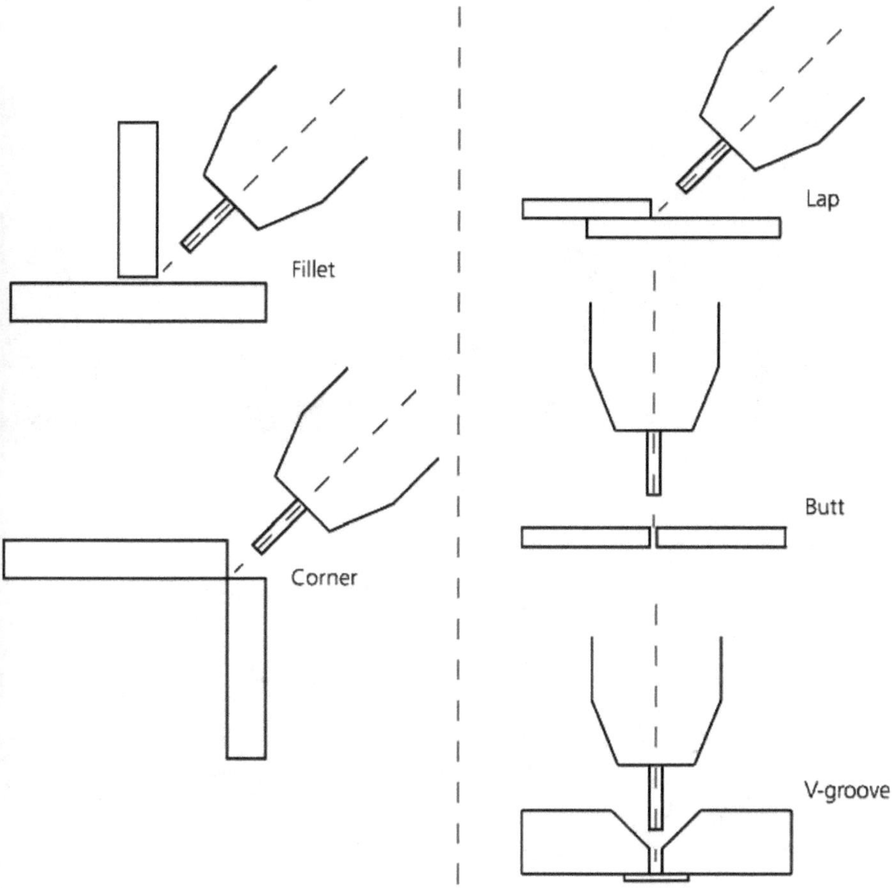

Figure 3.4. Typical standard joint types. *Left column*: fillet and corner joint. *Right column*: lap, butt and V-groove joint [3]

A laser-based seam tracker is typically mounted on the weld torch and has the weld joint in its field of view some distance ahead in the weld path direction, see Figure 3.3. This means that the robot must use one degree of freedom to keep the sensor in alignment with the weld joint during welding or, alternatively, use a separate motion so that the sensor can rotate around the weld torch to maintain the alignment relationship between the sensor and the weld torch. It should also be noted that the seam tracking sensor must measure and deliver target positions of the weld torch continuously and that these must be time stamped and stored in a buffer for later use by the robot controller.

In order to use the data from the laser scanner, weld joint features must be extracted from the image and a target position must be determined which is stored in the input buffer to the robot controller. The feature extraction algorithm is dependent on the weld joint to detect and is defined beforehand. Examples of different weld joint types are shown in Figure 3.4.

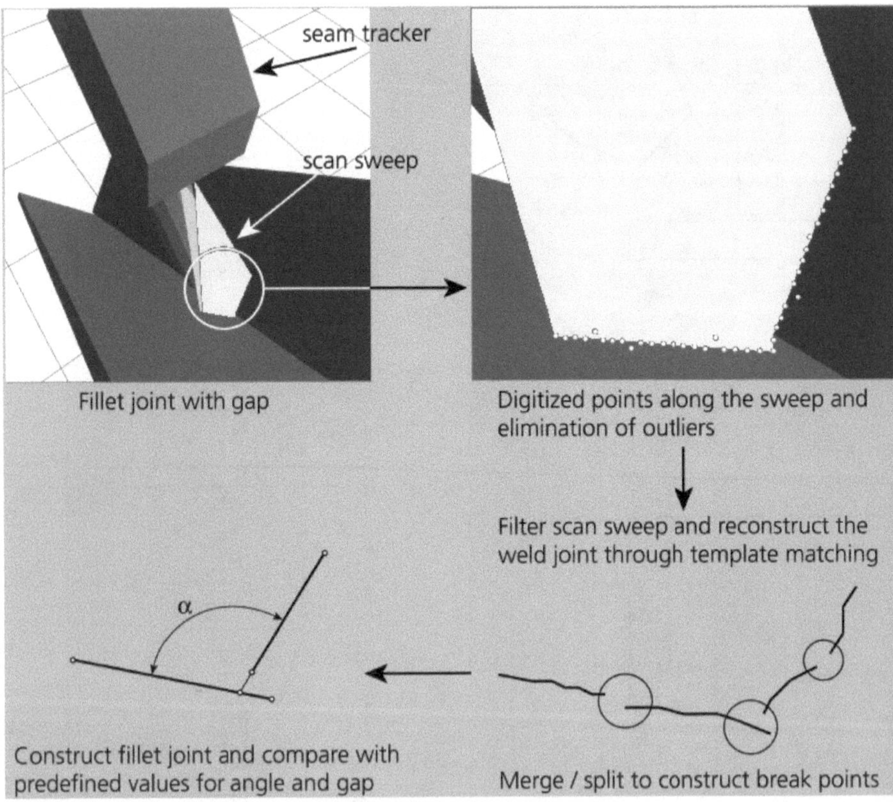

Figure 3.5. Example of the steps of feature extraction of the segmentation process: (1) outlier elimination from the scan, (2) line segmentation generation based on the specific joint template, (3) join the line segments, and (4) validate against templates and tolerances [3]

The feature recognition process includes the following basic tasks: (i) identification and elimination of outliers, (ii) contour generation of the weld profile and generation of line segments based on predefined templates, (iii) merging line segments, and (iv) validation of joint parameters so that they are within predefined tolerances and match the joint template, see Figure 3.5.

From a control point of view, seam tracking is usually performed with full compensation of the position error measured. Seam tracking is typically only performed using a nominal path. The nominal path is the assumed trajectory of the weld joint and during tracking, the robot controller receives new target positions from the sensor and the robot controller overrides the nominal path by changing the position of the TCP while keeping a constant orientation. This has some benefits and drawbacks. The benefits are that, given a nominal path, it is rather straightforward to verify the ability of the robot to follow the path with some minor changes while keeping the orientation constant. This means that issues related to joint limits, singularities and possible collisions are minimized. Drawbacks are that the user must define and program a nominal path.

If, instead, the robot is only instructed where to start and where to end, it must be able to follow the path measured by the sensor on the fly. This put some additional requirements on the robot system as it must be able to calculate the trajectory including both target positions and orientations of the weld torch. In doing so, the robot might easily get into control problems for the specific task that is related to close to singular areas, joint limit and collisions with the work-piece. However, the benefits high and more flexible control of seam tracking can be obtained using available sensors.

In order to use the sensor data and let the robot follow the measured and generated trajectory, the target data from the buffer of target positions must be filtered. A suggested method is to use about five target positions and generate a polynomial curve for x, y and z and a vector tangent at the current position. This is useful for several reasons: (i) a generalized description of the weld joint makes it easy to reuse the generated trajectory for calibration purposes, (ii) the generated trajectory can be used as a nominal path for subsequent weld passes, (iii) the vector tangent can be used for subsequent target positions even if target drop-outs occur such as at tack welds and (iv) the vector can be compared with the current trajectory vector and from that an optimal orientation of the weld torch can be calculated with respect to the weld joint as measured by the laser scanner.

Typical operating data of a laser scanner is a scan sweep frequency of 10-50 Hz. If we assume a welding speed of 20 mm/s, this means about one sweep per mm during welding. This is in most cases more than sufficient. However, new welding processes such as laser welding will increase the welding speed considerably and for high requirements careful analysis and trials must be made. The accuracy of the laser scanner is high and is better than 0.1 mm. However, it should be noted that single scan sweeps may generate outliers as the weld joint and the environment during welding generate severe disturbances.

From a practical point of view, laser scanners are accurate and robust sensors that meet most requirements within the welding process. However, they must be mounted on the weld torch and take up some space. They also put additional requirements on programming and positioning of the robot and the weld torch. Laser scanning sensors are also still relatively expensive and if alternative methods can be used like through-arc sensing as described in the next section, they are usually preferred.

3.2.2 Through-arc Sensing

Seam tracking using a weaving motion and the arc itself as the sensor, sometimes referred to as through-arc sensing, was introduced in the 1980s. The principle behind the method is to make use of the change in current when the distance between the contact tube and the work-piece varies. The underlying principle is relatively easy and cost-effective to use and is a common sensor for tracking methods in robotic welding based on gas metal arc welding and related processes, like flux-cored arc welding, submerged arc welding, *etc.* According to [4], the approximate relationship between arc voltage (U), arc current (I) and the contact tube to work-piece distance (l) is expressed by

$$U = \beta_1 I + \beta_2 + \beta_3 / I + \beta_4 l \qquad 3.1$$

where the constants β_1 - β_4 are dependent on factors like wire, gas and the characteristics of the welding power source. In most cases, the welding power source is set up to maintain a constant voltage and thereby a more stable welding process. Thus, when the value l varies, the arc current I will also change, mainly as a proportional change with opposite sign. This can be used in mechanized welding and specifically in robotized welding to perform a weaving motion during welding. When doing so for a weld joint as shown in Figure 3.6, the distance between the weld torch and the weld joint will vary during the weaving motion and so will the current. Hence, a longer contact tube to work-piece distance will result in a lower arc current than a shorter distance, given that all other parameters are the same. This can be utilized during an overlayed weaving motion, normally a sinus or triangular type of motion, but more complex motions also exist.

In practice, the current is measured using a *Hall Effect* sensor or a current shunt. A low-pass filter is used to depress noise from the signal. From a control point of view, sensor data can be analyzed continuously using template matching or only at the turning points using differential control. Due to the relatively low accuracy of the sensor data regarding precision, *etc.*, a differential control is in most cases sufficient. This should be seen in the context that the sensing principle needs a few mm weaving amplitude to be able to measure a reliable change and difference in arc current. It also indicates that achievable tolerance are restricted for those cases which can accept a weaving motion, which produce a wider weld rather than a straight motion. However, in medium to thick plates this is usually not a problem (thicknesses above about 3 mm).

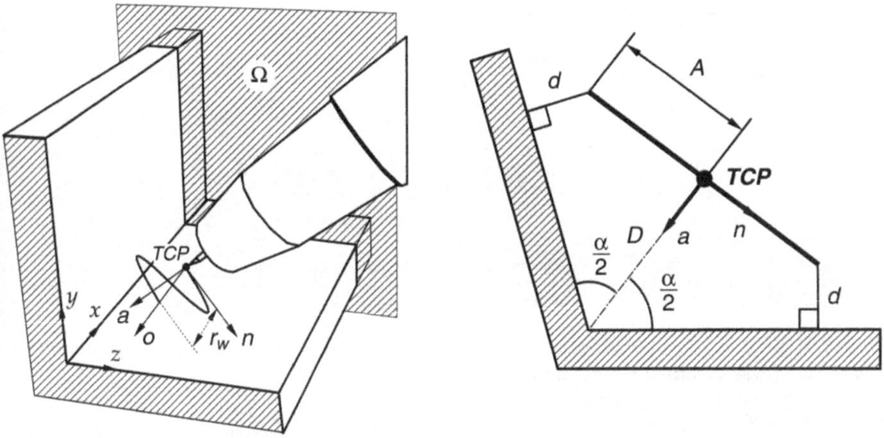

Figure 3.6. *Left*: definition of Tool Center Point (TCP) and weaving directions during through-arc sensing. *Right*: the optimal position for seam tracking in arc sensing [5]

Experiments based on the through-arc sensing principle indicate that it is possible to detect a variation in contact tube to work-piece distance of about ±0.25 mm. This shows that the method can be used for tracking with quite small amplitudes, in the order of 1 mm. However, it is important that the control of the motion is robust so that the robot does not lose the weld joint during tracking since there is no obvious way to get back on track. This is due to the fact that the weld joint can only be detected during weaving and welding and that the field of view is equal to the weaving motion of the welding gun. If no weld joint is detected, there is no information available on how to find the joint.

In practical implementations the tracking functionality is usually combined with a search function, specifically for the start-up phase, so that the robot starts welding with a weaving motion and, if no joint is found, gradually moves in a predefined direction perpendicular to the main nominal weld path. A search function can be implemented in basically two different way, (i) move the weld torch towards the respective plates one at a time until contact occurs, usually by measuring an electrical contact, and from that information calculate a starting position of the weld joint, or (ii) during welding define a start position for the weld and also define a direction of gradually moving the weld torch during weaving until it detects the weld and tracks the seam in a normal way.

The information acquired from through-arc sensing can be retrieved and used in basically two different ways, either continuous measurement of the current or measurements at the turning points of the weaving motion. If a height control is included, a measurement should take place at the center of the weld joint as well. Based on the measurement principles above, different control principles can be applied which are usually based on differential control and/or template matching of the signal. If template matching is used together with continuous measurement of the current, a more precise control can be made that can also take into account for non-symmetric shapes of the joint profile.

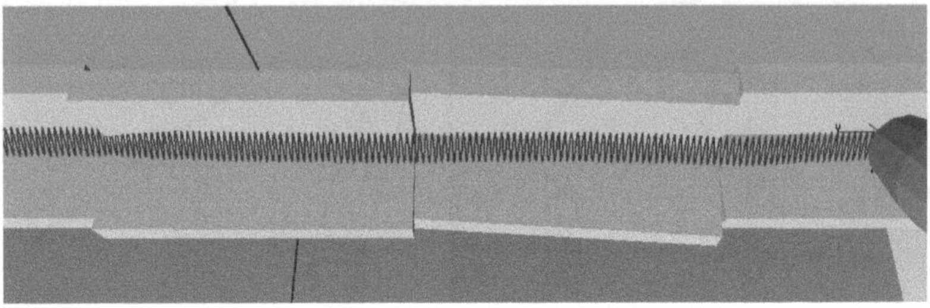

Figure 3.7. Example of the functionality of the through-arc seam tracking over segmented plates that deviate both sideways and in height [5]

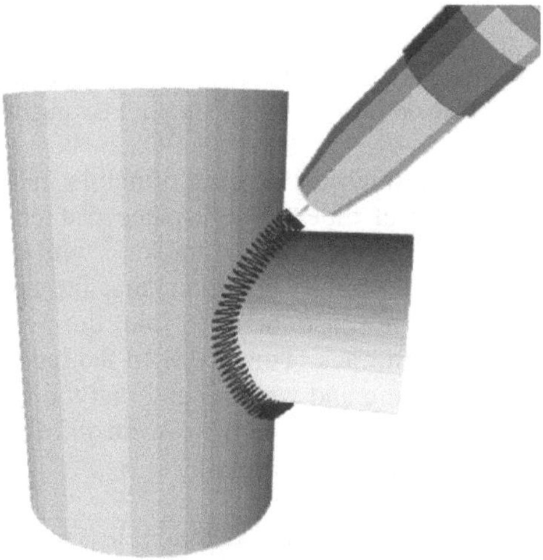

Figure 3.8. A T-pipe representing a type of work-piece that should benefit from a seam tracker which can compensate for both position and orientation changes [5]

As for seam tracking using laser scanners, a nominal path is used. In multi-pass welding the robot controller has a function to memorize the tracked path and use that as a template to make subsequent welds. Similar to tracking using lasers, tracking is performed to change the position of the weld torch so that it will be aligned with the weld joint. In general, no compensation will be made for orientation of the weld joint. However, a control scheme can be made that generates a polynomial and vector description of the weld path as described above and the sensor principle can be used to drive the robot with subsequent target positions during welding. A sample simulation of through-arc seam tracking is shown in Figure 3.7 which displays the principle of tracking over segmented plates which deviate both sideways and in height. In Figure 3.8, a work-piece (T-pipe) is shown as an example of a complex weld path that is difficult to generate a program for and thus, a sensor with a capability to both track and generate the weld path is a suitable technique.

3.3 Monitoring

The ability to monitor the weld quality automatically is important in order to reduce production costs and to assure and improve weld quality. An automatic detection system should be able to classify different weld defects such as porosity, metal spatter, irregular bead shape, excessive root reinforcement, incomplete penetration and burn-through.

Monitoring systems for weld parameters such as *ADM III*, *Arc guard*, *Analysator Hannover 10.1* and *Weldcheck* are commercially available [2],[6]. They all work in a similar way: voltage, current and other process signals are measured, presented and compared with preset nominal values. An alarm is triggered when any difference from the preset values exceeds a given threshold.

Thus, an important feature of monitoring is that it is done during welding and using data that exist during the welding process. To be able to make any judgment about the quality, reference data must be available including models or algorithms that describes and evaluate measured parameter.

An important task of any monitoring system which is used for quality assurance or quality control purposes is to be able to present the data with respect to quality measures as consistently as possible. This means that alarm thresholds defined must be correlated with real weld defects or relate to specifications defined in the WPS. An important aspect in this context is to understand that the welding process displays a more unstable situation when the data frequency of the readings are increased, and consequently, measurements of process parameters at lower frequencies, providing they display mean values, will display a more stable process.

The information within the WPS does not normally account for this but includes nominal operating data for different controllable parameters. Thus, part of the monitoring system for control purposes is to define alarm thresholds with respect to the WPS to maintain the process within nominal parameter limits and at the same time produce a weld at the defined quality and productivity levels.

The examples given here are limited to the detection of changes in the weld quality both automatically and on-line in spray GMAW when using signal processing methods. However, the method as such can in principle be applied to any welding method providing that knowledge exists about the stability criteria of the process and how to measure significant parameters related to the stability.

Gas Metal Arc Welding (GMAW) is widely used in welding applications because of the specific advantages it offers such as reduced spatter and smoother bead appearance as compared with submerged arc welding. There are two stable metal transfer modes in GMAW: short-circuit metal transfer at low arc voltage; and spray metal transfer at high voltage and high current.

In the field of GMAW of steel, both physical analysis of the welding process [7]-[11] and statistical analysis of real welding signals have been performed [12]-[21]. However, the problem of classifying the weld with respect to quality is still in focus for research and is an important area to produce efficient control systems which include, if not all, the most important parameters and how these affect the quality and productivity, and the proper definition of the corresponding WPS for control purposes.

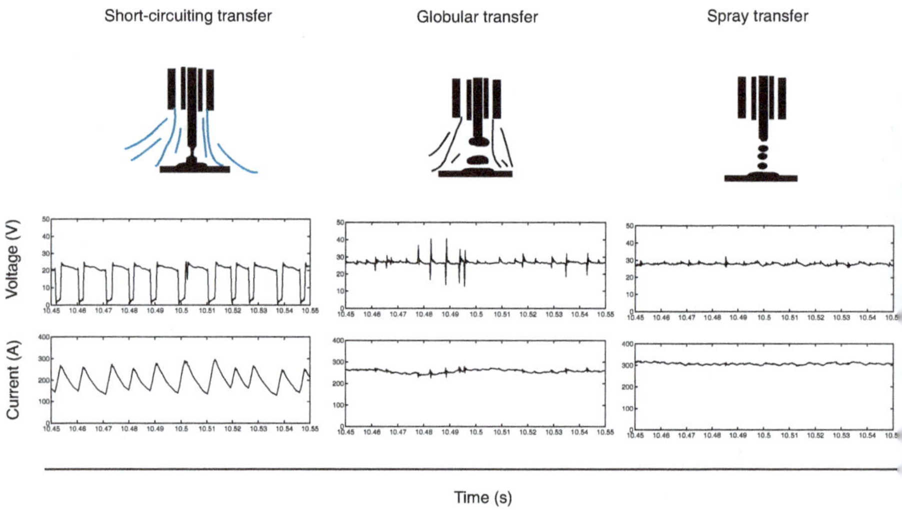

Figure 3.9. Weld voltage and current waveforms for different metal transfer modes

The GMAW process has three basic modes in which metal is transferred from the electrode tip to the work-piece, Figure 3.9. The modes can be classified as short circuiting transfer, globular transfer and spray transfer. In short-circuiting transfer, the metal is deposited during a short circuiting of the weld wire which normally lasts about 10ms. In globular transfer the droplet diameter is larger than the wire and is considered as an unwanted transfer mode. Spray transfer is a transfer mode where the drops are smaller than the weld wire. The type of metal transfer mode is a function of the weld current according to both the static force balance theory and the pinch instability theory [22]. As the mean current increases, the metal transfer goes from stubbing, through short-circuits and globular to spray transfer mode, see Figure 3.10.

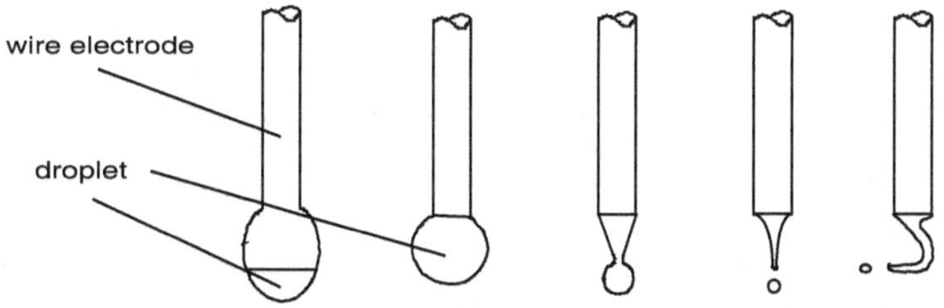

Figure 3.10. Successive transfer modes of metal transfer in GMA welding with increasing mean current (*left to right*) [23]

For the three main metal transfer modes there is a correlation between the voltage and current envelope waveforms and modes of metal transfer. The variation in the current and voltage waveform is reduced when moving from short-circuit to spray transfer and can be used to classify the transfer. To weld in short-circuiting GMAW mode, the open circuit voltage and the electrode wire-feed rate is set to a low value and in spray transfer mode, the open circuit voltage and the wire feed rate is set to a high value.

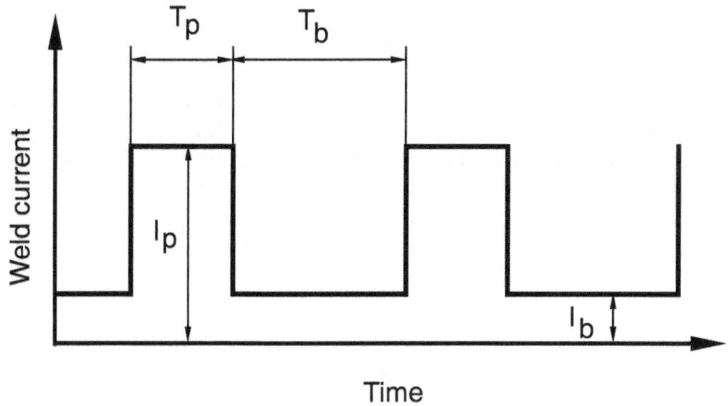

Figure 3.11. A schematic illustration of weld current and related parameters in pulsed GMA welding. T_p and T_b denote peak pulse time and background pulse time respectively, and I_p and I_b denote the peak current and background current respectively

3.4 Pulsed GMAW

In pulsed GMAW the amplitude of the current alternates between two levels, see Figure 3.11. The advantage of this method is that the mean current, and thus the average heat input to the work-piece, is lower than in spray GMAW. Thanks to the smaller heat transfer, it is possible to weld thinner plates with pulsed GMAW than with spray GMAW and at the same time maintain a high metal deposition rate.

To limit the heat input to the work-piece, the current is low during part of the current cycle. This part of the cycle is denoted "background pulse time", and is represented by T_b. The current during this part of the cycle comprises background current and is represented by I_b. It must exceed a specified critical value in order to obtain a stable arc, *i.e.* the arc must neither go randomly over the work-piece nor be extinguished [24][25]. This lower current limit is defined by an empirical constant K_1. During peak pulse time, T_p, the current is high. The electrode is molten, and a droplet is detached and transferred to the work-piece. The main force for detaching a droplet and transferring it is the electromagnetic force induced by the peak current, I_p. The gravity of the droplet mass plays a minor role. The criterion for detachment of one droplet per pulse is governed by the relationship

$$I_p^n \cdot T_p = K_2 \qquad 3.2$$

where K_2 is a constant depending on the material, and $n \approx 2$ [25]. To achieve a better result, droplets should be of the spray type. If droplets grow larger than the electrode diameter, a globular metal transfer results, leading to greater probability of short-circuiting, spatter, uneven weld bead and other fusion defects. The approximate droplet volume is given by

$$D_V \approx K_3 A \,(I_p T_p + I_b T_b) \qquad 3.3$$

where K_3 is a constant and A is the electrode cross-sectional area. Wire feed rate, W_f should match the burn-off rate W_b, so that a constant arc length can be maintained. This is important to avoid burn-backs and stubbing-in, which can cause defective welds. The mean current, I_m, is expressed by

$$I_m = \frac{I_p \cdot T_p + I_b \cdot T_b}{T_b + T_p} \qquad 3.4$$

and the wire burn-off rate is expressed by

$$W_b = K_4 \cdot I_m + K_5 \cdot I_m^2 \cdot l_e \qquad 3.5$$

where K_4 and K_5 are empirical constants for given materials and sizes. The first term describes the melting due to arc heat of the electrode tip and the second term describes the joule heating of the electrode stick-out (l_e) by the welding current flowing between the weld table and the electrode wire tip.

3.4.1 Synergic Control

The peak current of the current source used during pulsed GMAW can be current-controlled. This means that a preset current value will be given independently of the impedance of the welding process. Some welding sources have an option preventing the electrode from getting stuck during short circuitings by disconnecting the current control when the peak voltage decreases the certain preset value. The current can thereby increase until the short-circuit is broken.

To select the optimal welding parameters (W_f, I_p, I_b, T_p and T_b) by trial and error is a time-consuming procedure. Most manufacturers of welding sources thus use synergic control to simplify the parameter setting. With this method the operator simply chooses one parameter, *e.g.* the wire feed rate. The remaining parameters are adjusted automatically to their optimal values according to certain optimization criteria. The criteria to be met are [25]:

1. Arc stability: background current must exceed a minimum limit for stable arcing.
2. Metal transfer: spray type metal transfer must be produced.
3. Arc length: a constant arc length must be maintained.

The arc stability is maintained if a minimum background current is kept. It is also important to meet other conditions like clean metal surfaces and proper shielding gas flow during welding.

The metal transfer to obtain spray mode can be arranged by modulating the current in different ways and still keep the mean current at a certain value. The pulse width can be constant while varying the pulse repetition rate or the pulse repetition can be kept constant while varying the pulse width. The amplitude of the current pulse can also be varied to maintain a constant pulse size.

The arc length (l_a) can be estimated by measuring the weld voltage. The voltage over the electrode stick-out is small compared to the arc voltage and can in general be omitted for most monitoring and control purposes. The arc voltage can be expressed as

$$U_a = \beta_1 \cdot I + \beta_2 + \frac{\beta_3}{I} + \beta_4 \cdot l_a \qquad 3.6$$

where β_1, β_2, β_3 and β_4 are empirical constants. If we can assume that the time constant of the process is much less than the time constant of the pulse duration and fixed current, the arc voltage can be expressed as

$$U_a = K_5 + \beta_4 \cdot l_a \qquad 3.7$$

where K_5 is a function of the current. A change in arc length will cause a change in arc voltage and also a change in weld voltage which makes it possible to estimate the arc length by measuring the weld voltage. As the arc length can be estimated, it is possible to build a control system that keeps the arc length constant.

The burn-off rate can be controlled by changing the mean current. This can be used to adjust wire burn-off rate to match the wire feed rate and a constant arc length can be maintained.

The parameters to measure and monitor in pulsed GMAW are the pre-set values of the weld process (I_p, I_b, U_p, U_b, T_p, T_b). The monitoring system must be able to detect unacceptable variations in the welding features that may generate weld faults.

3.5 Short-circuit GMAW

In short-circuiting welding the mean current, and thus the average heat input to the electrode wire and the work-piece, is lower than in spray arc GMAW. Due to the smaller heat transfer, short-circuiting welding makes it possible to weld thinner plates than with spray arc welding. The metal deposition rate and the joint penetration are, however, less than in spray and pulsed welding. In contrast with pulsed GMAW, the power source in short-circuiting and spray GMAW is voltage-controlled. This means that the power source continuously changes its current in order to maintain the voltage constant at the output.

The short-circuiting cycle begins with an arc that is struck between the electrode wire tip and the work-piece. The wire electrode melts and a small droplet is formed at the electrode tip. This part of the cycle is denoted "arc time" and represented by T_a.

During the short-circuit time, T_s, the droplet at the end of the electrode touches the weld pool and a bridge of liquid metal is formed between the electrode and the weld pool. At this stage the arc will extinguish, the voltage will decrease to almost zero volts and the current will increase to its maximum value. Due to the high short-circuit current, necking of the liquid bridge starts to occur, ending in rupture of the liquid bridge. The molten metal is then transferred from the electrode tip to the weld pool by the force of the surface tension of the weld pool, the gravitational force and electromagnetic pinch force (induced by the current). After the droplet is detached from the electrode and transferred to the work-piece the arc is reestablished and the cycle starts over again.

The short-circuiting transfer mode has two degenerated modes of metal transfer, the short-term short circuiting, meaning that the short circuits last for durations less than 1-1.5 ms; and stubbing of the electrode in the weld pool, with the result that the droplet growth cannot occur in the normal way. Stubbing-in of the electrode, short-term short-circuits and globular metal transfer are considered to be unstable, while short-circuiting and spray transfer are considered to be a stable process.

In order to produce weld joints with uniform weld quality, it is desirable that the welding process is stable. This means that the metal transfer from the electrode wire to the work-piece should be as regular as possible. Experiments indicate that in the short-circuiting mode, optimal process stability occurs when the short-circuit frequency is equal to the oscillation frequency of the weld pool [26],[27]. The experiments also show that the weld pool oscillation frequency is mainly determined by the width of the weld.

When the short-circuit frequency and oscillation frequency of the weld pool are unsynchronized, the oscillating weld pool surface will fail to contact the growing droplet at the electrode tip at regular intervals. Hence, larger variation in arc time, short-circuiting time and transferred droplet mass will occur, and as a result there will be a lower process stability. When the short-circuit rate is synchronized with the weld pool oscillations, optimal process stability is obtained.

From the physics of the short-circuit GMAW process it follows that a natural choice of feature to monitor is the mean and standard deviation of the short-circuit frequency. Other features have been suggested in the available literature and include for example the use of an arc stability index which is based on the standard deviation of short-circuit time, the arcing time, the average short-circuit current and the average arc current. The specific features used may be dependent on the specific algorithm and the welding condition. However, studies made have shown that the mean and standard deviation of the arc time, the short-circuit time, the peak current, the short-circuit rate and the standard deviation of the short-circuit rate *etc.* are less effective features than the variance of the weld voltage in order to detect a defect weld [1].

3.6 Spray GMAW

In spray GMAW the metal drops are smaller or of the same size as the diameter of the electrode wire. This was the original type of metal transfer used when GMAW was initially developed. Due to the large heat input to the work-piece when welding in spray transfer mode, the weld joint produced has good penetration, which often is desirable. On the other hand, the considerable heat input also creates a large weld pool which may be difficult to control. Thus normal spray transfer is limited to flat and horizontal positions and cannot be used to weld thin materials.

The standard type of power source for GMAW has voltage amplitude characteristics which are not constant but which drop slightly with increasing current. The advantage of such power sources is that they have a tendency to maintain a constant arc length. This has been termed "*self-adjusting arc*" [23]. When the contact tube to work-piece distance is increased the arc length is also increased. The disturbance in the form of arc length enlargement results in a change in the working point along the power source slope. Arc voltage increases in this way by a change in voltage, at the same time as the mean current decreases. It should be noted here that a small variation in arc voltage results in substantial current swing. Reduction of the current leads to a decrease in the wire burn-off rate resulting in an automatic adjustment of the arc length towards its stable operating point. Decreasing the arc length has the opposite effect. This indicates that a change in contact-tube to work-piece distance is reflected almost entirely in the change of the electrode extension.

The metal transfer mode is not the only property of the welding process that is influenced by the weld current. Penetration width, depth, burn-off rate, arc stability and arc rooting are all also influenced by the weld current. It is thus desirable that the mean current is constant and stable. The current is not so easy to control since small variations in arc voltage results in a large current swing. From the discussion on the physics of the GMAW process it follows that the mean amplitude of the weld current is likely to be a suitable feature for detection of changes in weld quality.

3.7 Fault Detection Using Monitoring

The strength and appearance of the weld joint are important properties when welding two metal pieces together. In GMAW the strength of the weld joint is mainly determined by the bead width and weld penetration. Parameters in GMAW which affect the weld penetration and bead geometry are for example:

- Welding current
- Arc voltage
- Welding speed
- Wire feed rate
- Electrode extension
- Electrode composition and diameter
- Shielded gas composition and flow rate

Considerable skill and experience are needed to select the optimal welding parameters for each weld application. Deviation from the optimal settings of the parameters may change the weld joint geometry and cause deterioration in the weld joint strength. Furthermore, deviation from the optimal settings will also increase the probability of such weld defects as cracks, porosity, undercuts, microfissures, *etc.*, which will further decrease the strength of the weld joint.

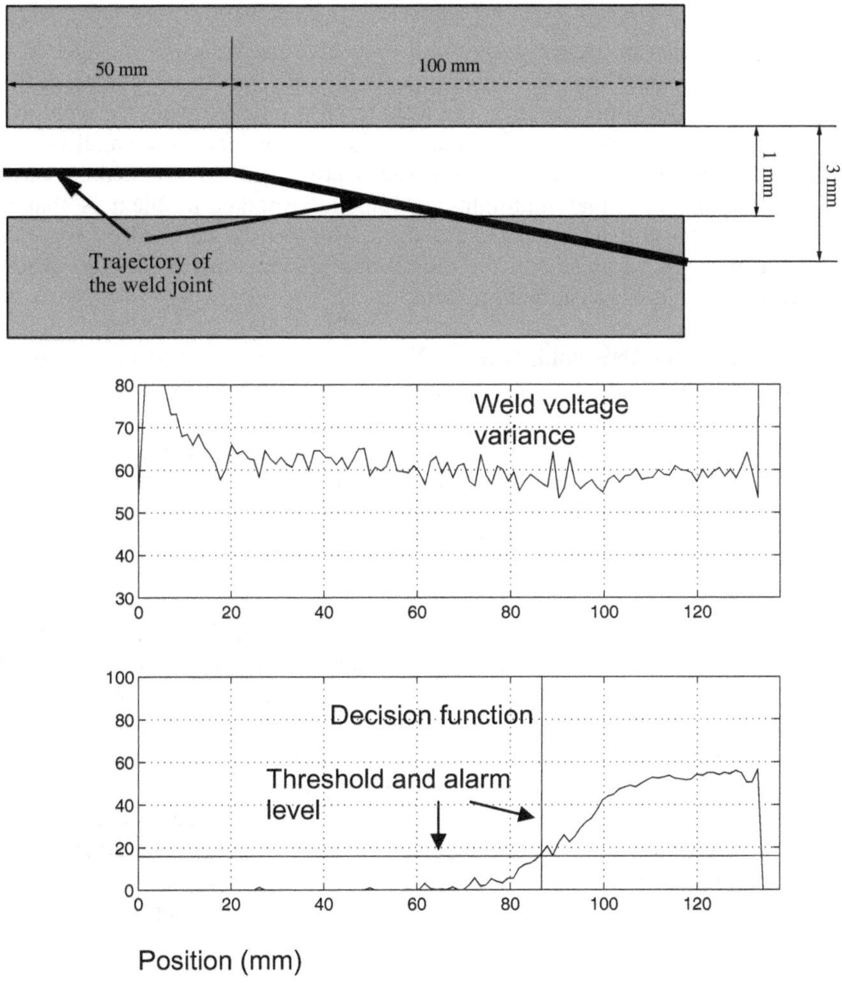

Figure 3.12. Butt joint with out of joint weld path. The weld voltage variance shows only a slight change as the weld errors occur. A sophisticated monitoring feature will however provide a robust alarm detection, in this case a sequential probability ratio test (SPRT) [1]

In practice, however, the optimal weld parameters for a specified weld joint are difficult to achieve. An experienced welder uses instead visual inspection of the arc stability and surface profile of the weld joint in order to select the proper weld parameters. For monitoring and quality control purposes, different strategies and use of the technology can be applied. Nominal parameter settings are usually defined prior to welding and any deviation from these values should be considered as a potential reason for decreased quality. However, it is also important to understand the acceptable tolerances of the parameter settings and how different parameters affect the weld. Thus, the task of the monitoring system is to measure welding parameters and extract features that indicate the quality of the weld with respect to preset values and provide these as a basis for further control actions.

A number of defects in the weld joint will only give rise to minor changes in the weld voltage and current or some other process parameters. As an example, a butt-weld is welded; after some distance, the weld torch is not following the weld joint, see Figure 3.12. By only measuring the welding parameters, only small changes can be observed. Modern monitoring systems are however designed to extract features from the data that highlight weld faults. Another problem is that too sensitive detection will generate false alarms. This can be handled by regarding random fluctuations as noise and the monitoring system must be able to observe and classify such signals as a random variable.

To select a detection threshold, two conflicting requirements must be considered. First, the threshold should be low enough to ensure that the probability of detection is not too small. Second, the threshold should be high enough to ensure that the false alarm probability will not be too large. For example, in practice, the false alarm probability must be low when welding hundreds of meters of tube. The welding process must not be stopped every meter due to false alarms. A simple detection algorithm can be realized by simply comparing the actual variance of the weld voltage (AC power level) with a particular preset level referring to a normal welding condition. In this way the difference between the preset level and the actual level is measured. A small difference will indicate a normal welding condition, while a large difference indicates a fault condition. For short welds the situation is the opposite and the system must be more sensitive to process disturbances which can produce quality problems. A general problem is that there is a run-in phase of the weld process which is difficult to monitor with respect to quality control. This means that a smaller proportion of short welds can be monitored with respect to quality and also for counteracting control. In practice, different strategies can be used, specifically if the same welding operation is made many times. In such cases template matching techniques can be used. If the weld is for low volume or one-off production, generic methods must be applied.

3.8 Design of a Monitoring System for Quality Control

The task of a monitoring system for quality control is to extract some features of the welding process, filter these as necessary and make use of them to detect changes in the process. The system consists of sensors, a signal conditioning unit, a feature extraction algorithm and a fault detection algorithm, see Figure 3.13. For control purpose, specific control algorithms must be included which should include the capability to map the monitored data with respect to quality and productivity specifications.

Sensors provide information about the system being monitored like the weld current, the arc voltage, the weld speed, the electrode extension, the wire feed rate and the shielded gas flow. They are all representing properties that affect the weld.

All parameters except the electrode extension can be measured directly, but the electrode extension can be estimated from the weld current and the weld voltage.

The analog signal conditioning unit carries out low-pass filtering of the analog signals from the sensors in order to prevent aliasing after the A/D converter. As the environment around the measurements of the weld process can be highly disturbed by electrical noise, the filter has the additional task of suppressing such noise. The bandwidth required for the monitoring system is, however, mainly determined by the short-circuit time, which is in the order of milliseconds, or about 500 Hz. The rise time of the whole system should not exceed 10 % of the short-circuit time and thus, the bandwidth needed should be in the order of 5 kHz. This makes it possible to extract features like peak pulse time, background pulse time and short- circuit time with adequate accuracy. Furthermore, high voltage transients caused by spatter or short-circuitings can also be measured.

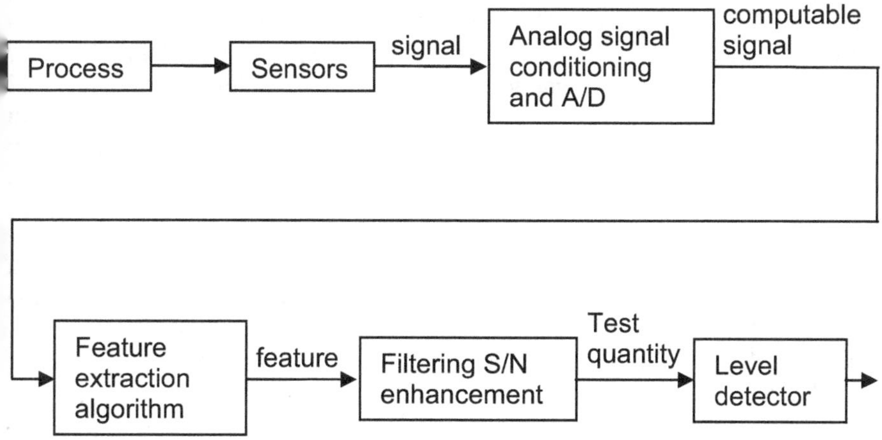

Figure 3.13. Block diagram for a monitoring system

Most modern commercial monitoring systems for GMAW divide the process into time segments or windows and the sample mean for each segment is estimated. These features are filtered and compared with a predefined threshold. As a result of the mean value calculation, information about higher frequencies is lost since the mean value of the signal has gone through a low-pass filtering process.

An appropriate feature which retains information about higher frequencies is the variance. This is due to the fact that the variance (or the area below the curve of the power spectral density) is an AC power estimate. A change in variance due to disturbance is an indication of changes in the power spectral density.

Another feature which may be relevant for monitoring purposes is the peak of the weld voltage spectrum. When welding in either the spray arc or short-circuiting mode, the spectrum peak changes when going from the normal welding condition to a fault welding condition.

When using an extended monitoring system, time information can also be used. Some of the features refer to the parameters of pulsed GMAW like peak and background current, peak and background voltage, and peak and background pulse time. In short-circuiting welding the features are typically short-circuit time, arc time, peak current or arc voltage.

However, for a simple monitoring system the primary feature to be monitored when welding in the spray mode is the mean of the weld current. The variance of the weld voltage can be used when welding in short-circuiting or pulsed mode. Also, a short-circuiting detector algorithm can be useful when welding in pulsed mode.

The fault detection algorithm provides variance reduction of the monitored features as well as a limit detector. The principle of applying a variance reduction technique, *i.e.* the filtering of the features, is that it results in increasing detection reliability by increasing the signal-to-noise ratio for the resulting test quantity, *i.e.* the filtered features. The algorithms developed and used (and partly exemplified here) should be applied to each feature extracted from the signals measured. It is important to note that data from the normal welding condition is assumed to be available in order to train or validate the algorithms.

It should therefore be noted that a monitoring system for quality control purposes should not only measure the defined welding data and compare these with the nominal, but also be able to calculate the variance of measured data to retain information about higher frequencies. To do this, algorithms must be adapted for fast calculations and response. This is needed since the bandwidth of the monitoring system basically is dependent and selected based on features from the process to be monitored. Thus, an indication of the frequency of pulses in short-circuit mode or pulsed GMAW is in the order of 100 Hz or 10ms. During monitoring, a number of pulses or drop transfers should be monitored as some could be outliers with respect to process stability criteria without affecting the quality of the weld. A typical value could be 500 data samples representing about five pulses or drop transfers. But given this, to define a threshold for correcting measures of the process is a delicate decision. If one such quality estimate indicates a change in the process stability, we might want to wait with any control actions until we have the estimate from the next 500 data samples. The reasons for this are simply that (i) such short disturbances do occur, (ii) more information is in general needed to get an indication how to stabilize the process, and (iii) in most cases, the weld processes do have some quite different characteristics with respect to time; many parameters and phenomena are truly rapid in nature and need some special attention to measure and monitor these, but the resulting weld is in general somewhat forgiving if we are able to react on a disturbance and get the process back to a stable arc again within a short time. How short this time is, is in principle dependent on the size of the weld pool. It is of course better to react quickly, but for most cases, control measures should be taken within 5 mm weld, which usually is in the order of 0.5 s. It should also be noted that a process disturbance in real welding usually comes gradually and in such cases, it is important for the monitoring system to detect this and make a smooth adjustment of the weld parameters.

From a practical point of view there are many reasons to customize and tune the control of the process. In most cases, any changes that may affect the quality gradually increase, like a change in the gap. If the change is abrupt, we might not be able to compensate for quality by control actions but have to stop the process. However, in such cases, we can accept waiting for some time with the decision from the monitoring system. This indicates that it is a matter of compromise between measured and detected irregularities on the one hand and resulting faults as defined in the WPS on the other, including trade-offs between quality and productivity. Unwanted stops during a weld may introduce quality concerns as well as decrease productivity.

Thus, the use of monitoring techniques should be an integrated part of the robot welding system including the definition of *Welding Procedure Specifications*. The algorithms used to monitor the process will depend on the process and what critical fault to detect. Another issue is the sensitivity to errors, which must be handled with an in-depth understanding of the purpose of the monitoring system and what is does.

3.9 Monitoring System Development – An Example

Some examples from experiments will be described here to show the benefits in quality control of the welding operation performed by the robot [1]. For this purpose, it is important to apply a method that is fast and robust with respect to detection time and number of false alarms. The setting of these parameters should, for each case, be possible to adapt according to the requirements as there is generally a trade-off between quality *vs* productivity.

3.9.1 Short-circuiting GMAW

In order to produce weld joints of uniform weld quality, it is desirable that the weld process is stable. Experiments have shown that in the short-circuit mode, optimal process stability occurs when the short-circuit frequency equals the oscillation frequency of the weld pool. Optimal process stability can also be described to have:

- A maximum short-circuit rate (Number/s)
- A minimum standard deviation of the short-circuit rate

- A minimum mass transferred per short-circuit
- A minimum spatter loss

A deviation from the optimal welding conditions is assumed to lead to a higher probability of spatter, uneven weld-bead and other fusion effects. To detect this, it is shown that monitoring of the weld voltage and using the variance of the voltage is a robust feature. Other features could also be used like short-circuiting rate and arc time, but they are less robust to detect defective welds.

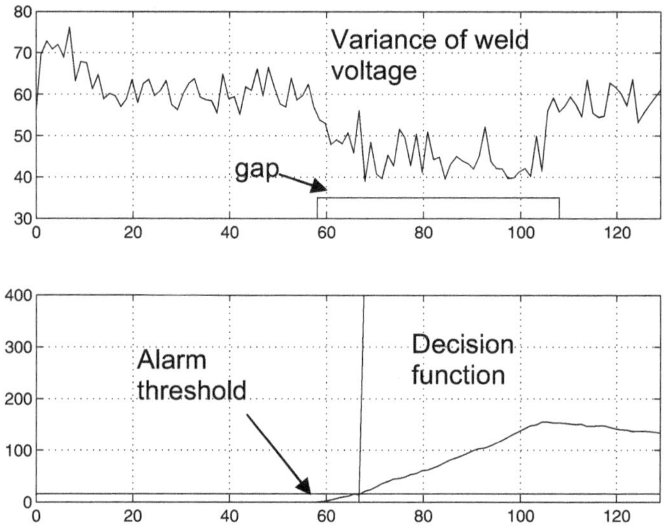

Figure 3.14. Sample T-joint with a step disturbance as illustrated with a gap. The decision function will set the alarm above the threshold level. The values of the x-axis represents millimeters along the weld joint [1]

The variance of the weld voltage is measured over 1024 samples and calculated. For each sample section an alarm decision function is accumulated if the variance decreases below a predefined value. The alarm value of the decision function will indicate accumulated successive variance measurements below the alarm threshold and the defined threshold will thus reduce the number of false alarms, see Figure 3.14. As mentioned before, this is a robust method to deal with the trade-off between quality and productivity and such alarm limits must be defined case by case.

3.9.2 Spray GMAW

Due to the self-regulated weld voltage and the much higher swing in weld current, the mean of the amplitude of the weld current is a suitable parameter for detection of changes in the weld quality when welding in spray arc mode. The weld current influences not only the metal transfer mode, but also penetration, width, depth, burn-off rate, arc stability and arc rooting. It is therefore desirable that the mean current is constant and stable.

By selecting suitable statistical detection algorithms a change in the weld condition can be detected that produces decreased quality. This can be combined with a decision function that accumulates over time and produces a weld error alarm after a preset threshold value, defined case by case.

3.10 Discussion

In general, applying sensors to measure parameters at the welding process is a difficult task as the process environment is a challenge to apply sensors in with high temperature, liquid metal, high current, spatter, *etc.* Due to this, it is common that parameters which we can observe by sensors are not the same as we want to control. Thus, a simple feed-back control is often not adequate or possible to obtain. Instead, models must be developed that map the observable parameters to proper actions within a model based control concept that focuses on relevant issues defined within a welding specification procedure. Here, productivity and quality measures should be defined together with nominal data to produce the weld (weld joint data, weld data for the welding power source and for the robot). Therefore, to utilize fully the information from sensors, a model based control should be applied together with welding procedure specifications that are defined with robotized welding and model based control in mind.

Seam tracking sensors are the most commonly used sensor systems in robotic welding. In general, two types of seam trackers exist, laser triangulation based scanners and through arc trackers which need a weaving motion of the weld torch during welding. Although laser based sensors have a better performance with respect to accuracy and resolution, the choice is not obvious in a real case. The through arc sensor works only during welding and can only measure anything through the arc and hence, the operating "scan" is equal to the weaving amplitude. This is critical for the run-in sequence of a weld using the sensor. However, the sensor detects the weld right at the weld arc with reasonable accuracy, although not all joints can be detected in a proper way compared to a laser sensor which is mounted in front of the welding torch and thus takes up some space and also requires being in line with the weld joint during welding. In addition to this rather scattered picture, the price of the two most frequently used seam-tracking sensors are quite different; the through-arc sensor is a low cost sensor while the laser sensor is to be considered as rather expensive. Thus, in reality the choice is dependent on many interacting factors within the specification of the task to be performed.

Monitoring systems are for the most part used in production where high quality is required. They provide in general the necessary means to work in real time with quality assurance and, if needed, stop the welding operation if the weld is found to be defective with respect to predefined specification and/or nominal operating parameters. Methods also exist to use the data from monitoring systems to control the weld process in real time through a model based control scheme. However, such control needs to be tuned for every case and is not in general use today.

3.11 References

[1] Adolfsson S, "Automatic Quality Monitoring in GMA Welding using Signal Processing Methods", PhD Thesis, Lund University, Dept. of Mechanical Engineering, 1998.

[2] Ågren B, "Sensor Integration for Robotic Arc Welding", PhD Thesis, Lund University, Dept. of Mechanical Engineering, 1995

[3] Cederberg P, Olsson M and Bolmsjö G, "Virtual triangulation sensor development, behavior simulation and CAR integration applied to robotic arc-welding", Journal of Intelligent and Robotic Systems, 35(4):365-379.

[4] Cook GE, *et al*, "Electric arc sensing for robot positioning control. Robotic Welding", pages 181-216. Editor: J. D. Lane. IFS Publications Ltd, UK, Springer-Verlag, 1987.

[5] Fridenfalk M, "Development of Intelligent Robot System Based on Sensor Control", PhD Thesis, Lund University, Dept. of Mechanical Engineering, 2003.

[6] Blakely PJ, "Developments in monitoring systems for resistance and arc welding", In Proceedings of the International Conference on Automated Welding Systems in Manufacturing. Gateshead, North East UK, November 1992. Woodhead Publishing Ltd., paper 40.

[7] Kisselevski F, Shvydkii N and Dolinenko V, "Mathematical simulation of scanning arc in MIG-welding", In Proceedings of the International Conference on Automation and Robotisation in Welding and Allied Processes, pages 146-150, Strasbourg, France, 1985.

[8] Amson JC, "Lorentz force in the molten tip of an arc electrode", British Journal of Applied Physics, 16:1169-1179, 1965.

[9] Allum CJ, "Metal transfer in arc welding as a varicose instability: I. Varicose instabilities in a current-carrying liquid cylinder with surface change", British Journal of Applied Physics, 18(7): 1447-1468, 1985.

[10] Hermans MJM, Spikes MP and den Ouden G, "Characteristic features of the short circuiting arc welding process", Welding Review International, 12(2):80-86, November 1993.

[11] Rehfeldt D and Schmitz T, "Investigation and measurement of weld pool oscillation in GMAW", Technical Report, IIW-Doc 212-882-95, 1995.

[12] Lucas W, Ahmed N and Hutt G, "Process stability in MIG welding" The Welding Institute Research Bulletin, pages 329-336, October 1983.

[13] Leino K, Nikkola A and Vartiainen K, "Prediction of weld defects using welding condition data", Research Report 264, Technical Research Center of Finland, Espoo, 1984.

[14] Gupta SR, Gupta PC and Rehfeldt D, "Process stability and spatter generation during dip transfer in MAG welding", Welding Review, pages 232-241, November 1988.

[15] Mita T, Sakabe A and Yokoo T, "Quantitative estimates of arc stability for CO_2 gas shielded arc welding", Welding International, 12(2):152-159, 1988.

[16] Liu S and Siewert T, "Metal transfer in gas metal arc welding: Droplet rate", Welding Journal, 68(2):52-58, 1989.

[17] Xie M, "Quality Assurance and Control for Robotic GMA Welding", PhD Thesis, Lund University, Dept. of Mechanical Engineering, 1992.
[18] Ogunbiyi B and Norrish A, "GMAW metal transfer and arc stability assessment using monitoring indices", In Proceedings of sixth International Conference on Computer Technology in Welding, Lanaken, Belgium, 9-12 June 1996. TWI, Abington Publishing.
[19] Adolfsson S, Bahrami A and Claesson I, "Quality monitoring in robotized welding using sequential probability ratio test", In Proceedings of TENCO '96. Digital Signal Processing Applications, volume 2, pages 635-640, Perth, Western Australia, 27-29 November 1996, IEEE.
[20] Shinoda T, Nishikawa H and Shimitzu T, "The development of data processing algorithms and assessment of arc stability as affected by titanium content of GMAW wires during metal transfer", In Proceedings of sixth International Conference on Computer Technology in Welding, Lanaken, Belgium, 9-12June 1996. TWI, Abington Publishing.
[21] Adolfsson S, *et al*, A"utomatic quality monitoring in robotized GMA welding using a repeated sequential probability ratio test method", International Journal for the Joining of Materials, 9(1):2-8, 1997.
[22] Kim JA and Eager NM, "Analysis of metal transfer in gas metal arc welding", Welding Journal Research Supplement, 76(6):269-278, 1991]
[23] Lancaster JF, "The Physics of Welding", Pergamon Press, Oxford, UK, 1986
[24] Amin M, "Synergic pulse MIG welding. Metal Construction", 13(6):349-353, 1981.
[25] Amin M, "Pulse current parameters for arc stability and controlled metal transfer in arc welding", Metal Construction, 15(5):272-278, 1983.
[26] Hermans MJM, Spikes MP and den Ouden G, "A stability criterion for short circuiting gas metal arc welding", In Proceedings of the International Conference on the Joining of Materials, JOM-8, pages 112-119, Helsingor, Denmark, May 1997.
[27] Rehfeldt D and Schmitz T, "Investigation and measurement of weld pool oscillation in GMAW", IIW-doc 212-882-95, 1995.

4

Robotic Welding: System Issues

4.1 Introduction

Robotic welding research deals with the relevant technical and scientific aspects involved in the task of reproducing the work of the experienced and skilled human welder. Welding was for a long time a task performed only by humans, being a craft that combines skill with art and science. Automating welding is therefore a very difficult and demanding objective, because of the required adaptive behavior of the automatic system.

It can be considered that any welding operation is constituted by three very different phases [1],[2]:

1. **Preparation phase:** where the welding operator sets up the parts to be welded, the welding apparatus (power source, robot, robot program, *etc.*) and the welding parameters. The type of gas and the type of wire are also selected in this phase. If any CAD/CAM or other offline programming facility is used then a robot welding pre-program is available and should be placed on-line. This aspect is very important since currently most of the welding pieces are designed using CAD software. Consequently, that software should be used to generate robot programs that could work as starting points for the welding tasks, needing only minor tuning due to calibration [3]. That may be done easily by the welding operator just by performing selected on-line simulations of the process [1]-[3], calibrating in this way the robot program that should then be ready for production.
2. **Welding phase:** considering a manual welding operation, the welder acts by adjusting the process variables just by continuously observing the welding operation and the correspondent results. If automatic equipment is used to perform the welding operation, then the same capabilities must be present, *i.e.*, the system should be able to maintain the torch orientation while following the desired trajectory (that may be different from the

planned one), perform seam tracking and change welding parameters in real-time. With those capabilities available the system should be capable of emulating the adaptive behavior showed by the manual welder.

3. **Analysis phase:** this is normally a post-welding phase where the manual welder, or welding operator in the case of an automatic welding system, examines the obtained welds and decides if they are acceptable or if changes are needed in the two previous phases. In the particular case of an automatic system, this phase can be performed automatically, or by means of user input using specific software interfaces. When advance sensors are used, like laser 3D cameras, this phase can be executed on-line during the welding phase. This is particularly interesting since evaluation of welding quality on-line may influence the ongoing welding process.

Consequently, when designing a fully automated robotic welding system all the above welding phases must be considered as a way to achieve a good performance and welding quality. The following sections detail some of the relevant problems, namely: modeling and control the welding process, system interfaces and programming environments.

4.1 Modeling the Welding Process

Modeling the welding process is basically a theoretical problem (a physics problem mainly) and a technological problem, *i.e.*, understanding the welding process requires theoretical studies but also extensive experimentation to obtain the governing models. Several of the most interesting welding processes were explained in Chapter 2, giving practical guidelines about the relationships between the variables and the parameters that characterize the welding process. Part of the current knowledge on welding is empirical and based on detailed experimentation, which focuses on technological aspects. Consequently, the strategy used in this book was to present the most interesting welding processes from a robotics and automation point of view, focusing on the technological characteristics and automatic system requirements. The physics of the process is briefly introduced and the reader referred to other technical publications, as a way to identify the process parameters relevant for each welding process.

4.1.1 Definition and Detection of the Process Parameters

To design a welding robotic system the first step is to identify the process related parameters, *i.e.*, the parameters that should be controlled in a way to obtain the desired quality, also defined by a set of accepted characteristics (Figure 4.1). The process related input parameters can be classified into three different categories:

1. **Primary inputs:** variables that can be modified on-line during the welding process. Taking as example the GMAW process, the primary

welding parameters are the *voltage*, the *wire feed rate*, and the *torch speed*. Technically, the *voltage* and the *wire feed rate* are analog signals commanded to the welding power source, and generated from the robot controller or process PLC. The *torch speed* is the desired speed commanded to the robot TCP for coordinated motion.

2. **Secondary inputs:** variables defined when the process is selected and before any welding service. Using again as example the GMAW process those parameters include the type or composition of the *shielding gas*, the *flow of gas* during the process, the *torch angle*, and the type and size of the *wire* to use.
3. **Fixed inputs:** parameters that are fixed and cannot be changed by the user. These parameters are usually an imposition of the selected welding process, of the current welding procedure or of the physical setup. Parameters of this type include the joint geometry, plate thickness, physical properties of the plate metal, *etc.*

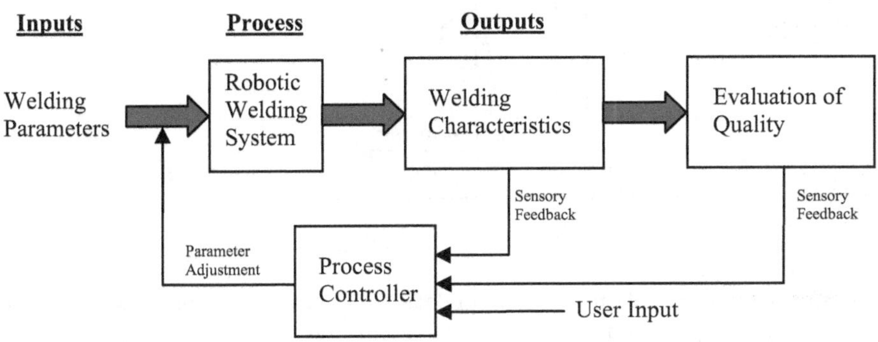

Figure 4.1. Overview of a welding control system

All these parameters must be handled carefully, namely the correct preparation of the setup and the selection of the secondary inputs are fundamental to control the primary inputs efficiently.

Another important set of parameters are the output parameters. Those parameters characterize the weld and are used to evaluate its quality. In a general way, there are two types of output parameters: geometrical and metallurgical.

Geometrical parameters result from the process mass balance, and basically define the way how the transferred metal fills the welding joint. Consequently, the basic parameters used to classify an acceptable weld are the *penetration*, the *bead width*, the *bead height* and the *cross-sectional area*, if we consider a V-Groove weld, and the *penetration* and *the length of both legs* for a fillet weld (Figure 4.2). The *penetration* is a very important parameter to evaluate the quality of the weld, because it is related to the way the weld metal combined with the base metal during the welding process. Unfortunately, it is very difficult to control the *penetration* during the welding process since there is no way to measure it on-line.

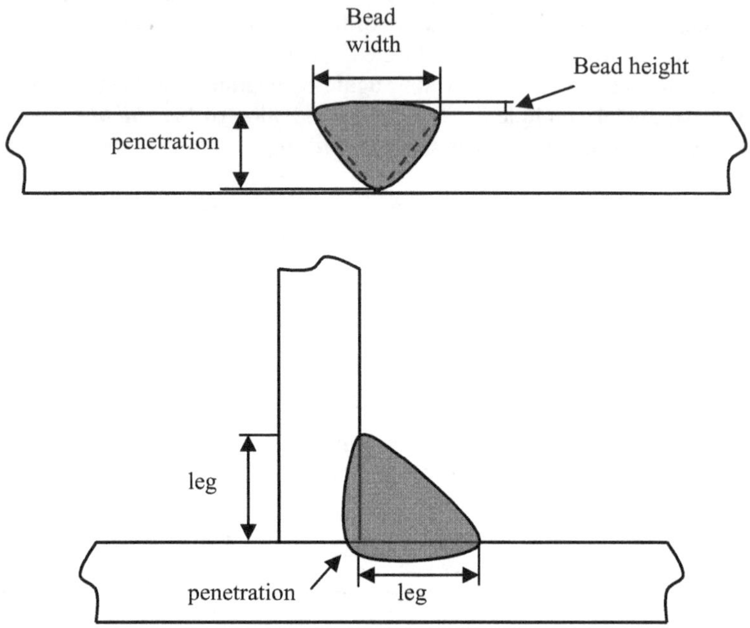

Figure 4.2. V-groove and fillet weld geometrical parameters

Sensors are used in robotic welding to detect and measure the process features and parameters [4]-[7], namely the joint geometry and the weld pool geometry and location, used for on-line control of the welding process. Nevertheless, sensors are also used to perform weld inspection and quality evaluation.

The first basic thing to achieve with a fully automatic robotic welding system is the capability to follow precisely the joint to be welded [3]-[11]. This is because the welding quality depends very much on the welding pool position apart from its geometry. A pre-programmed path cannot be obtained with the desired precision, since deviations from the programmed path are likely due to deficient path definition, but also due to material plate deficiencies and to the effect of heating the plates. Consequently, an on-line joint recognition and seam tracking system must be available. Several techniques have been used for joint detection and seam tracking, namely for welding robotic systems. Using the arc characteristics exploiting the proportional relationship between the welding current and the distance from the electrode to the work-piece, as proposed by Cook [12], was one of the first approaches. All technicians working in welding are familiar with the weaving techniques used to obtain the joint geometrical profile. In fact, the only thing that is needed is a current sensor and a comparison system: setting the reference as the current reading for a perfectly centered torch on the actual welding situation, the center of the welding joint can be obtained just be weaving the arc along the joint and reading the current signal [13]. When the torch is centered the difference between the two signals is zero (Figure 4.3), and the signal can be used to position the robot carrying the torch.

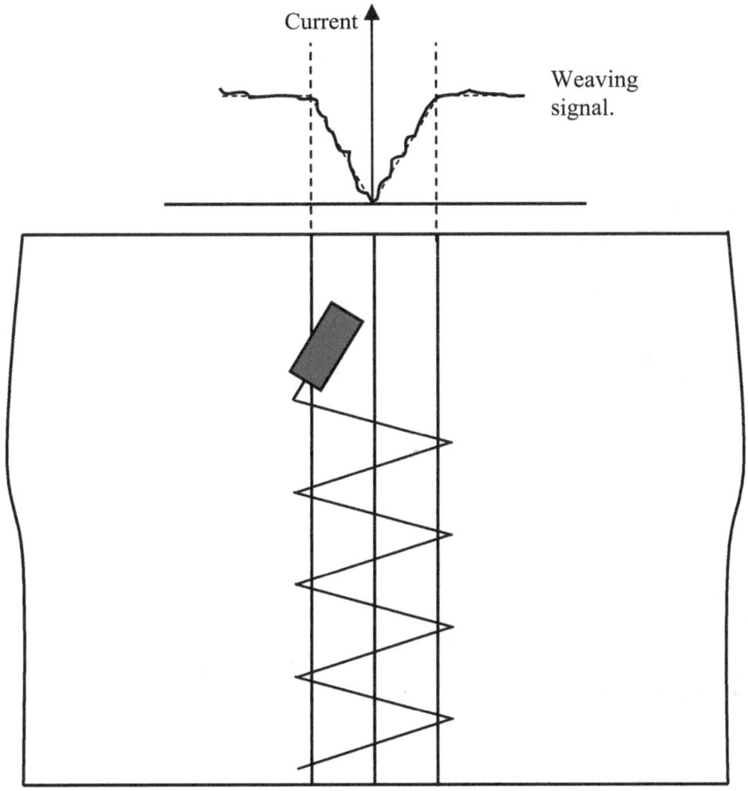

Figure 4.3. Using the current signal to find the joint center position

This approach is slow, does not work well with all types of joints and requires weaving motions which may not be desirable or possible for certain welds. Nevertheless, this is a commercially available system.

The utilization of vision systems permits more accurate results [4],[5],[12]. Many researchers tried to use CCD cameras to obtain the joint image and compute the track to follow and even evaluate quality of the weld [6]. Those approaches suffered from CCD saturation due to the light generated by the arc, and interference due to the electromagnetic field also generated by the electric arc, which did not make them ideal for on-line seam tracking. Nevertheless, they could be used for joint detection, for weld pool detection and for quality control, with the selection of an appropriate narrow band filter and/or optimizing the placement of the camera in such a way to avoid the arc light and electromagnetic interference. Since these solutions are not robust when several welding processes are considered, the most used approach with CCD cameras is to make a teaching pass before the welding process is actually initiated. Theoretically this is a good solution since a good reading of the welding seam can be recorded and used to guide the system during the welding process. The drawbacks of this approach are the reduction of arc-on time and the insensitivity to deviations, even if small, of the welding seam that can happen due to the extremely high temperatures characteristic of the welding process and deficiencies on the material of the plates to be welded [13]-[19].

The laser based 3D cameras have been used successfully for joint detection, seam tracking and weld inspection. These cameras work in a very simple way, based on the principle of laser triangulation. A low power laser source is used to generate a laser beam that is projected onto the surface of the joint to weld. The reflected light is picked up by a lens that feeds the imaging system composed usually of a CCD or CMOS sensor. The laser reflected signals are extracted using filters and image processing software, which is a simple task since the laser signal has a very precise wavelength and power (Figure 4.4).

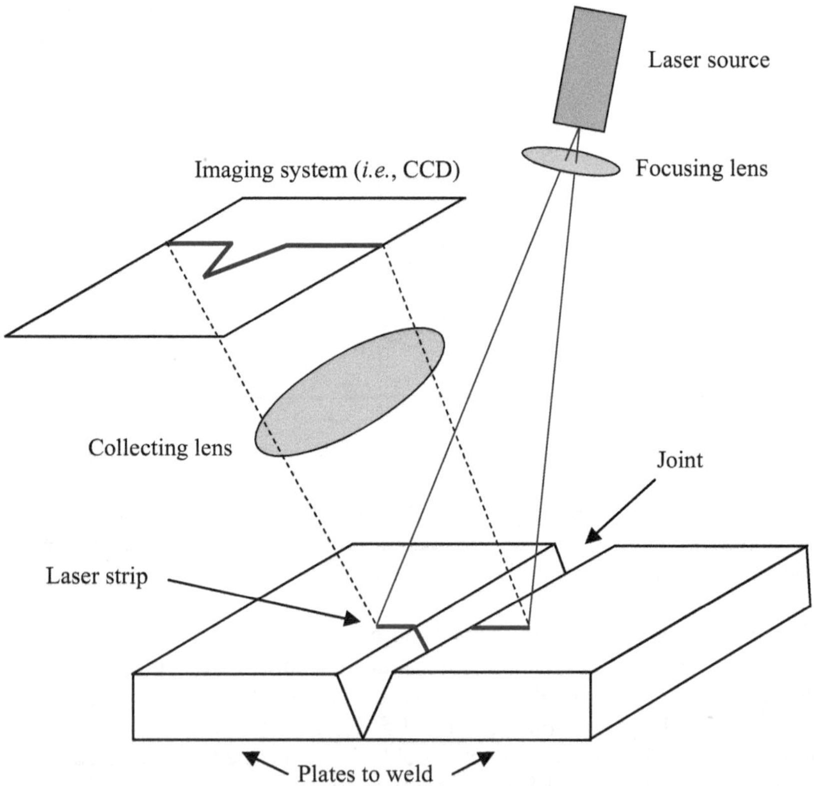

Figure 4.4. Explanation of the laser vision principle

In fact these laser cameras and related processing hardware and software, with some customization to the selected application [20], are very useful for evaluating most of the geometric parameters besides the already mentioned joint detection and seam tracking features. Since they are available with powerful APIs for general use, with standard interfaces for robot controllers and current computer hardware, this type of sensors constitute a powerful tool for robotic welding.

Another very important and challenging parameter is the penetration. Basically, a good weld has constant penetration along the weld path, and consequently the welding system should be capable of keeping that goal despite possible variations in the joint geometry. If full penetration is achieved then there are some methods to observe the penetration on-line, but in cases of partial penetration there are no means to monitor its evolution. Several methods were designed to measure the penetration [21] when full penetration is achieved, but most of them require back-face bead measurements requiring access to the back of the work-piece, which isn't always possible. Measuring front-face bead geometrical characteristics, along with the weld bead temperature provides the means to estimate the penetration. This means a good understanding of the welding process behavior, so that a precise model, correlating two dimensional measurements of the weld pools with the three dimensional shape of the same weld pool, can be written and used for on-line penetration control. Using ultrasonic techniques is also possible [21] as an alternative, since a full model description is difficult to obtain and, although desirable, has not been achieved yet.

Metallurgical aspects are also important for the welding quality, since they determine important mechanical characteristics like hardness, soundness, strength and residual stresses. Those very important mechanical parameters are not easy to measure on-line and are a consequence of several mechanisms. Nevertheless, they all result from the heat generated in the welding process. And since the welding process is basically based on heat, the following is needed to guarantee an acceptable weld:

1. A certain peak temperature is needed to achieve a good metal fusion and penetration.
2. A roughly uniform temperature distribution, centered in the weld joint, is required to achieve a constant weld.
3. Acceptable cooling rates, compatible with the required metallurgical characteristics of the final work-piece, are also needed.

All these requirements focus on the need to monitor all the thermal events of the welding process, which adds to the other geometrical measurements requirements that a successful robotic welding system must implement.

4.2 Control of the Welding Process

To control effectively the welding process the automatic robotic system should adapt to actual conditions, like the human welder does, and be able to move effectively the welding torch and control the power source. Several implementations are reported in the literature, but here the survey is post-1990 since surveys reported in the 1980s were more or less consistent in saying that few robotic systems had adaptive capabilities, although several had seam tracking capabilities [21].

Basically there are three major needs to be able to mimic effectively the human welder with a automatic robotic system (Figure 4.5):

1. A knowledge base
2. Sensors and interfaces
3. Programmable and flexible control system facility

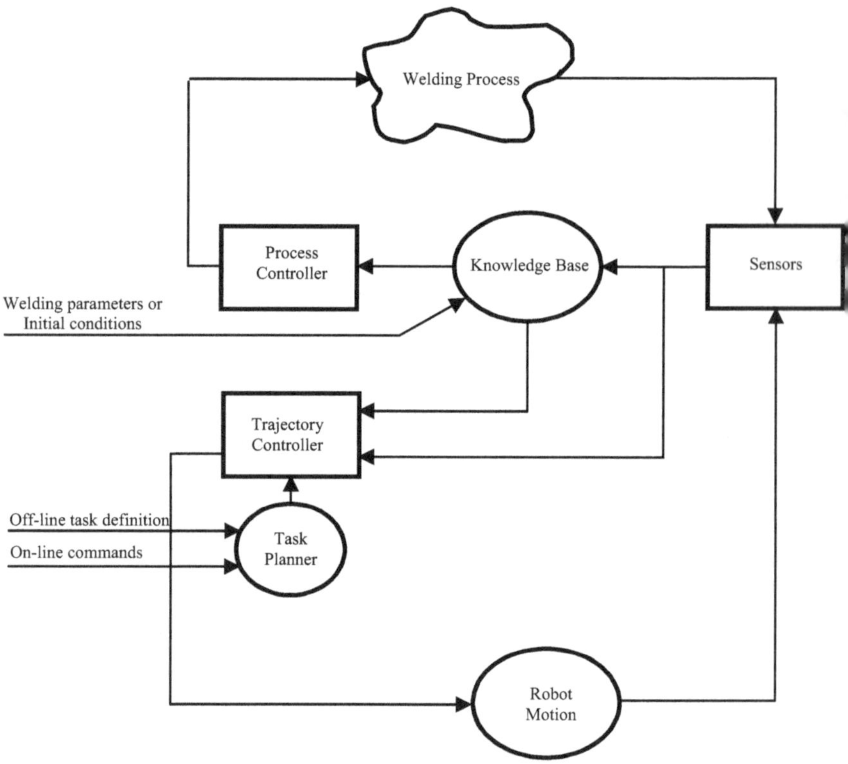

Figure 4.5. Basic scheme of a robotic welding control system

4.2.1 Knowledge Base

As already explained in Chapter 2, the welding process is very difficult to model which suggests that a rule base approach [8],[22],[23] would be advisable instead of an explicit mathematical model approach. That means exporting to a knowledge base the relevant information for a specific welding process and setup, so that the adaptive behavior could be still obtained. That can be done in several ways: using neural networks or fuzzy rule base approaches, simple lookup tables for the relevant parameters, *etc.* As always, simplicity is desirable and a solution that could grow with experience, just by adding new rules or more training, would be ideal for a physics process as complex as welding.

4.2.2 Sensors and Interfaces

The existence of smart sensors is fundamental to achieve a good solution for a robotic welding system that is necessarily distributed, *i.e.*, it is based in the distribution of functions through the components of the system as a policy for efficiency and organization. Since the most promising sensors and sensor techniques have already been presented in Chapter 3, the focus here is given to the interfaces and system architecture.

The basic components of a single cell robotic welding system include a robot manipulator, the robot controller, the welding torch, the welding power source, and the sensors adopted to monitor and sense the process parameters (Figure 4.6).

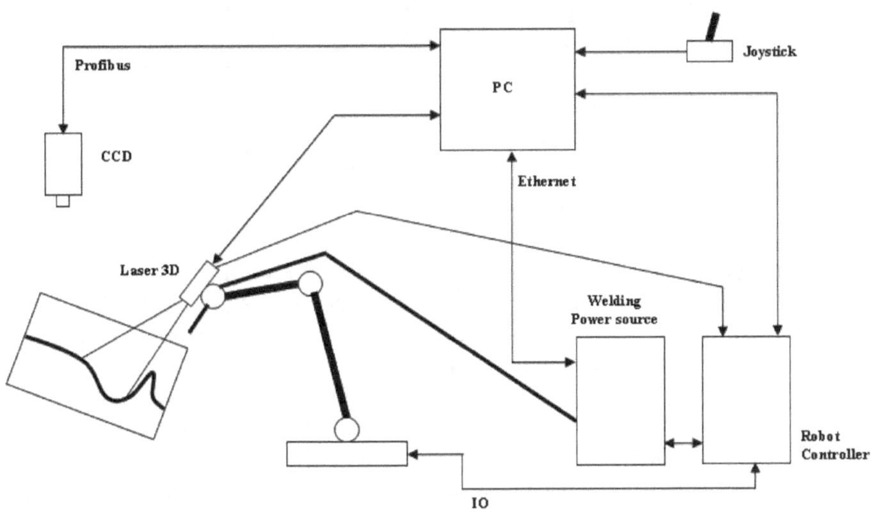

Figure 4.6. Single cell robotic welding system

Nevertheless, an industrial application may include several robots, and sensors, and welding power sources, which means that data networks must be present to connect all the parts of the system. These networks may range from TCP/IP Ethernet networks, used for example for high-level communication between computers and controllers, to fieldbuses, used for example to make high-frequency connections between sensors and controllers (Figure 4.7).

If the sensors are considered in particular, it should be emphasized that the perspective here is to reach intelligent sensors, *i.e.*, sensors that have a microprocessor, some way of dealing with Remote Procedure Calling (RPC) and asynchronous events. If that is available, then the sensor can be programmed to feed the necessary data at the specified sampling rate, when necessary, and fire programmed events when the relevant information is available. This means that the robotic welding systems must be ready to implement distributed software architectures and must be event driven systems, or at least must allow events to be used for obtaining asynchronous information and influence the system behavior.

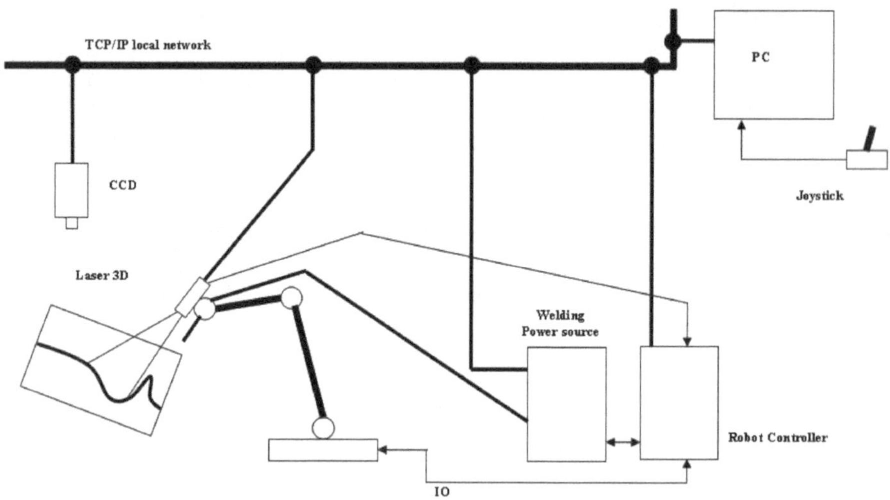

Figure 4.7. Networked robotic welding system: multi-cell

Let's consider for example a laser 3D camera, a sensor used for seam tracking but also for welding inspection [20]. For the seam tracking task all that is needed is the information from the sensor (available in the form of IO signals or a serial data interface), which can be used to guide the robot through the welding track, keeping the torch centered with the seam middle section. The sensor can be connected directly to the robot controller if the necessary signals and software interfaces are available, which is generally the case. Usually this is a low-level interface that acts on the robot motion controller and enables fast correction of the pre-programmed trajectories. Another solution could be a connection through the high-level programming interface, which is not desirable because it is high-level and necessarily slower. This should only be used for seam tracking if the low-level interface is not available in the robot controller.

Nevertheless, the above-mentioned high-level interface can be used for other non-intensive and lower frequency tasks, which could include inspection, monitoring surface scanning, *etc.* Being microprocessor based, sensors can be programmed to feed the controlling computer with the required data streams, respond to remote commands and fire events. That means that a collection of services, that includes event firing services, must be available in the sensor, along with a programming API that provides to the user means to explore all the sensors functionalities from the available network using some type of RPC (Figure 4.8).

The same applies to CCD cameras and generally to any microprocessor based and programmable sensor, *i.e.*, it should be possible to define all the sensor functionalities, and develop a set of functions and the interface mechanism that could enable the users to explore all the sensor features at the desired way and rate. When using a camera the primary data feed is not an image of the welding process, since generally those are large collections of data and require post-processing, which is not suitable for any real-time experiment. The vision systems should be able to acquire images, extract the relevant user-defined features just by processing the acquired image, and return them through the available interface. Those relevant features are usually simple things like areas, distances, *etc.*, *i.e.*, numbers that are very easy to transmit and don't overload the network with very long data streams.

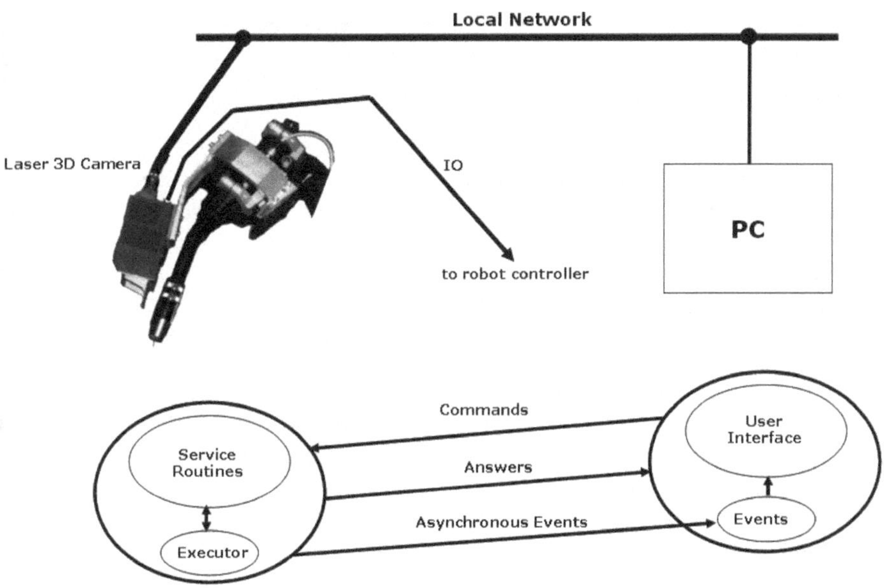

Figure 4.8. Using a programmable sensor

4.4 Programmable and Flexible Control Facility

To be able to use a robot manipulator, the required sensors and the necessary welding equipment, on a robotic welding application, it's necessary to outline the software architecture that could be used as the developing platform. The challenges posed by a robotic welding system are not different from the ones posed by a typical flexible manufacturing system based on robots. Consequently, the software architecture presented here was designed to be used with general robotic manufacturing cells that may include several types of equipment like robot manipulators, mobile robots, PLCs, CNC machines, vision systems and several types of sensors, *etc.* Usually these systems use different programming languages, even when the manufacturer is the same. It is then very difficult to make adjustments to the cell functionality, or adapt it to new requirements posed by the introduction of a new product or by changes introduced in existing products. Several research and technical efforts have been made to overcome these problems. Many of those efforts point to solutions that consider the development of general programming languages that could be used with any equipment, relying on individual interpreters to generate the specific code for any equipment. Nevertheless, recent research works show that it is desirable to have a flexible environment and still program each machine using its own language [24],[25]. The reason is simple: a general syntax means introducing generalizations and simplifications that tend to limit the potentiality of the equipment. Consequently, some parameterization is not used, special non-grouped functions are not used and the generated code takes always a uniform structure which may not be the best for all machines.

The idea presented here is significantly different, being an alternative for the solutions presented in the literature [27], and also for the software products truly distributed available on the market [28]-[30]. The basic idea is to define for each individual machine a collection of software functions that expose all its basic operational features. That objective requires local processing capabilities, availability of communication channels and support for the standard technologies used when implementing the services, installed on the individual machines. Since the vast majority of the current robotics and automation (R&A) equipment meets these requirements fully, this is not a serious limitation. It should also be stressed that the above-mentioned services are to be offered through a local network, on a distributed software framework based on the client-server model. Furthermore, using those services from the remote client computer to build controlling and inspection applications can be performed from any platform (UNIX, Linux, Win32-DCOM, *etc.*), using standard programming languages (C, C++, C#, Visual Basic, *etc.*).

Several approaches can be used and are currently available from various robot manufacturers, with specific details and implementations. Nevertheless, the following objectives are pursued by any of the above-mentioned software architectures:

1. Be able to represent the robot manipulator motion based on the kinematic and dynamic models, but also based on real-time data coming from the real robot. That can be done using available mathematical and graphical software packages, like Matlab® for example. This latest objective clearly indicates the need to access robot motion and status information in real-time from the mathematical package.
2. Be able to develop applications to explore remotely the entire installation (robot and welding application, for example) using standard programming languages (C, C++, C#, Visual Basic, *etc.*).
3. Be able to integrate and explore intelligent sensors used to obtain information from the process under control.
4. Enable users to explore the advanced programming capabilities of actual robot controllers, namely the local programming capabilities, based on a dedicated programming language complemented by extensive libraries of functions, and the optimized manipulation capabilities based on trajectory planning software that takes advantages of optimized kinematic and dynamic models.
5. Enable users to build flexible manufacturing cells, which leads to the possibility to explore the available industrial data network, the possibility to distribute software to the various components of the system, and the capacity to build remote software applications to control and monitor industrial manufacturing cells;
6. Develop advanced Human Machine Interface (HMI) solutions to operate with industrial systems, hiding from the users all the tricky details about implementation, allowing them to focus on the operational details, *i.e.*, to focus on how systems work and how they can be explored efficiently.
7. Provide ways that could allow developers to focus on the important things about the application they are building: the control algorithm, program functionality and HMI. All the details related with communications, sensor integration, *etc.*, should be hidden from the user.

Taking into consideration these objectives the following programming models are required:

1. **Client-server model**: there should be server code running on each cell equipment, namely on the robot controllers and coordinating PLCs, that could receive calls from the remote client computers, execute the commands and return the results;
2. **Remote Procedure Calls**: this is the most usual method used to implement communications between a client and a server on a distributed environment. The client makes a call to a non-local function and the selected RPC mechanism configures the call so that the proper computer, server program and function are addressed, adding the necessary network headers. The server program, running on the server machine, receives the call, executes the selected function and returns the results obtained to the client computer.

3. **Data sharing**: most of the services require data sharing, files and databases between the client and the server. Consequently, the mechanism provided by the RPC technology to implement data sharing must be used.

Another important thing to consider is the need to interface intelligent sensors with the system. The most easy and portable way to do that is to build software components that implement the methods, properties and data structures necessary to configure and use the sensor. Consequently, a technology to implement software components is also needed. The basic architecture presented in Figure 4.9 details all these requirements.

4.5 Application to Robot Manipulators

Actual industrial robot manipulators are controlled by advanced multiprocessor computer systems, based on standard parallel buses (VME, for example) or a serial internal communication mechanism (CAN, DeviceNet, *etc.*). Generally these robot control systems use a real-time operating system for low level interfaces, like RTOS or WxWorks, and a more friendly operating system with the sub-systems used for user interface (Win32 based OS, *etc.*). Robot controllers also provide local programming environments based on structured Pascal-like languages, along with a set of libraries that enable the user to build custom applications, interface with other machines and with operators, *etc.*, exploring fully the robot and controller facilities.

Any software architecture designed to operate with this type of machine should comply with current standards in terms of communication protocols, remote interfaces and software components. The reason is to avoid incompatibilities and excessive dependency with specific technologies that limit the users or force them into certain directions not representative.

The software presented in this book was designed to be used with robot manipulators (Figure 4.9) in distributed applications, and is divided into three main parts:

1. A set of functions that implement the robot-PC communication operations, including the access to the RPC services available in the robot controller. Those services include: variable access services, file management services, program management services, IO control, robot controller state services, *etc.*
2. A set of functions based on TCP/IP sockets that implement the same robot controller access described above. These functions were designed to operate with a TCP/IP server running as a task on the robot controller.
3. An OPC (OLE for Process Control) client component that implements calls to any OPC DA (Data Access) server [31]. The particular implementation used in this book works with ABB IRC5 OPC DA servers [31] to demonstrate the principle.

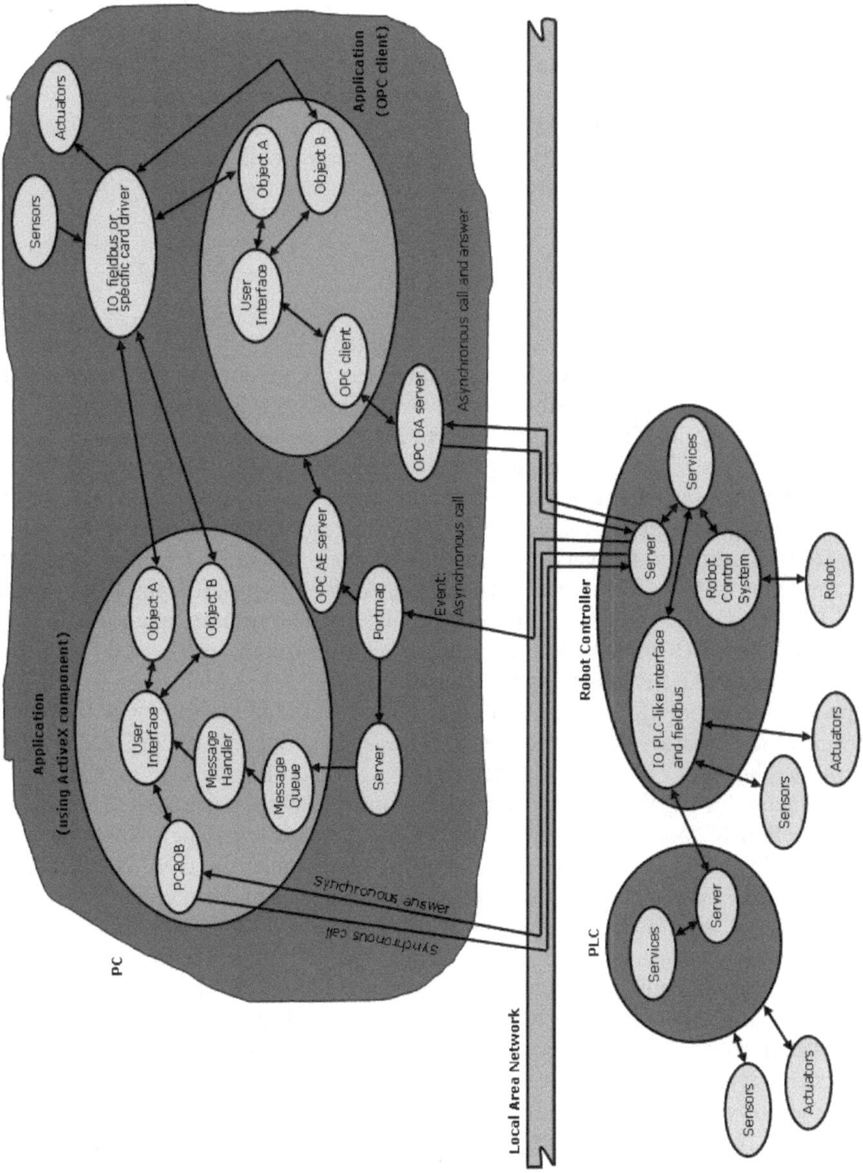

Figure 4.9. Software architecture used (depicting two approaches: using software components and OPC – OLE for Process Control [31])

Using any of the presented software interfaces it's possible to design a software strategy/architecture to implement robot controller access for data exchange, program control and robot controller status supervision. To particularize and demonstrate a few implementation specific details, two different technologies will be explored in the following sections: remote procedure calls (based on the SUN RPC 4.0 implementation [32]) and TCP/IP sockets (based on the Microsoft Winsock implementation [33]-[35]). The concepts will be kept as general as possible, using a specific robot controller when needed for clarity.

4.5.1 Using RPC – Remote Procedure Calls

Client-server software architecture was adopted [32],[33]. The robot controller software works as a server, exposing to the client a collection of RPC services that constitute its basic functionality (Figure 4.9). Those services, offered by the RPC servers running on the robot controller, include as already mentioned variable access services, files and programs management services, and robot status and controller state management and information services. To access those services the remote computer (client) issues properly parameterized remote procedures calls to the robot controller (server) through the network (Figure 4.9). Considering for example the S4CPLUS robot controller from ABB Robotics, it's possible to extend the RPC services available in the robot controller adding user functionality to the system. The ABB implementation is based on a messaging protocol developed by ABB called RAP (Remote Application Protocol) [38], which is an Application Specific Protocol (ASP) of the OSI application level. The messaging protocol RAP defines the necessary data structures and message syntax of the RPC calls used to explore the RPC services available in the controller. These services were implemented using the standard and open source RPC specification SUN RPC 4.0, a collection of tools developed by the SUN Microsystems Open Network Group (ONC) [32]. Consequently, to implement the client calls, the ONC SUN RPC 4.0 specification and tools were also used. This package includes a compiler (rpcgen), a portmaper and a few useful tools like rpcinfo [32]. The Microsoft RPC implementation uses another standard defined by Digital Corporation named OSF/DCE [37], which is not compatible with the SUN RPC standard. The package used to build the client software was a port to Windows NT/2000/XP [39], equivalent to the original version that was built to UNIX systems, although a few functions were slightly changed to suit better the needs without compromising compatibility with client and server programs developed with the SUN RPC package. The port was compiled using the Microsoft Visual C++ .NET 2003 compiler [42].

From all the RPC services available in the robot controller, the ones really needed to implement the software architecture depicted in Figure 4.9 are the variable access services. Nevertheless, calls to the other services were implemented for completeness. The procedure is simple and based on the XDR (Extended Data Representation) file obtained by defining the data structures, the service identification numbers and the service syntax specified by the RAP protocol. That file is compiled by the *rpcgen* tool, generating the basic calls and data structure prototypes necessary to invoke the RPC services available from the robot controller. The necessary code was added to each basic function and packed into an ActiveX software component named PCROBNET2003 [39]-[41]. The complete set of functions included in this object is listed in Table 4.1.

It should be noted that, although this software component was built using the DCOM/OLE/ActiveX object model, it does not run the Microsoft RPC implementation but instead the already mentioned SUN RPC 4.0 port to the Win32 API.

Table 4.1. Methods and properties of the software component PCROB NET2003 [40],[41]

Function	Brief description
open	Opens a communication line with a robot (RPC client)
close	Closes a communication line.
motor_on	Go to run state
motor_off	Go to standby state
prog_stop	Stop running program
prog_run	Start loaded program
prog_load	Load named program
prog_del	Delete loaded program
prog_set_mode	Set program mode
prog_get_mode	Read actual program mode
prog_prep	Prepare program to run (program counter to begin)
pgmstate	Get program controller state
ctlstate	Get controller state
oprstate	Get operational state
sysstate	Get system state
ctlvers	Get controller version
ctlid	Get controller ID
robpos	Get current robot position
read_xxxx	Read variable of type xxxx (there are calls for each type of variable defined in RAPID [43])
read_xdata	Read user defined variables
write_xxx	Write variable of type xxxx (there are calls for each type of variable defined in RAPID [43])
write_xdata	Write user defined variables
digin	Read digital input
digout	Set digital output
anain	Read analog input
anaout	Set analog output

To use a remote service the computer running the client application needs to make properly parameterized calls to the server computer, and receive the execution result. Two types of services may be considered: synchronous and asynchronous. The synchronous services return the execution result as the answer to the call.

Consequently, the general prototype of this type of calls is

short status call_service_i (**struct** parameters_i, **struct** answer_i)

where *status* returns the service error codes (zero if the service returns without errors, and a negative number identifying the error otherwise), *parameters_i* is the data structure containing the service parameters and *answer_i* is the data structure that returns the service execution results.

The asynchronous services, when activated, return answers/results asynchronously, *i.e.*, the remote system should also make remote procedure calls to the client system when the requested information becomes available or when the specified event occurs (system and controller state changes, robot program execution state change, IO and variable events, *etc.*). Those calls may be named events or spontaneous messages, and are remote procedure calls issued to all client computers that made the correspondent subscription, *i.e.*, made a call to the subscription service properly parameterized specifying the information wanted. To receive those calls any remote client must run RPC servers that implement a service to receive them (Figure 4.10). The option adopted was to have that server broadcasting registered messages inside the operating system, enabling all open applications to receive and process that information by filtering its message queue.

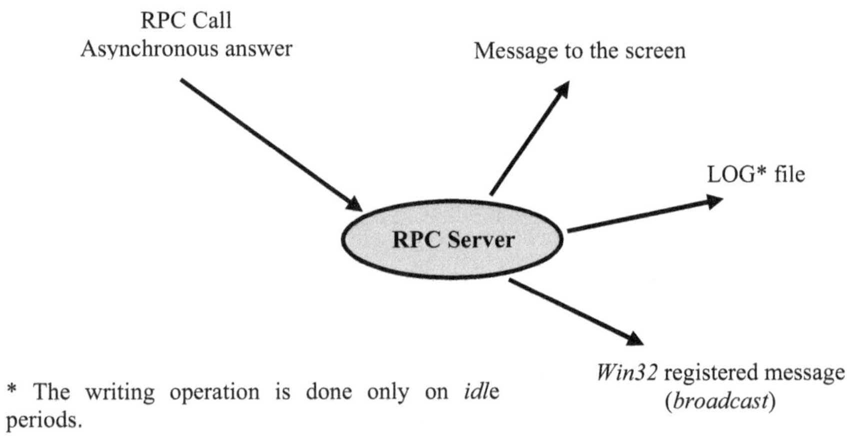

Figure 4.10. Functionality of the RPC server necessary to receive spontaneous messages

As mentioned already, the variable access services allow access to all types of variables defined in the robot controller. Using this service, and developing the robot controller software in a convenient way, it is possible to add new services to the system. In fact that possibility may be achieved by using a simple SWITCH-CASE-DO cycle driven by a variable controlled from the calling (client) remote computer:

```
switch (decision_1)
{
    case 0: call service_0; break;
    case 1: call service_1; break;
    case 2: call service_2; break;
    ...
    case n: call service_n; break;
}
```

This server runs on the robot controller, making the process of adding a new service a very simple task. The programmer should build the procedure (routine) that implements the new functionality, and include the call to that procedure in the server cycle by identifying it with specific number of the control variable.

In fact this is not far from what is done with any RPC server; the *svc_run* function, used in those programs is a SWITCH-CASE-DO cycle that implements calls to the functions requested by the remote client. With this type of structure it is straightforward to build complex and customer functions that can be offered to the remote client. Furthermore, with this approach it's still possible to use the advanced capabilities offered by the robot controller, namely the advanced functions designed to control and setup the robot motion and operation.

Examples exploring this facility are presented and discussed in Chapter 5.

4.5.2 Using TCP/IP Sockets

One of the most interesting ways to establish a network connection between computer systems is by using TCP/IP sockets. This is a standard client-server procedure, not dependent on the operating system technology used on any of the computer systems, which only requires the definition of a proper messaging syntax to be reliable and safe. The user defined messaging protocol should specify the commands and data structures adapted to the practical situation under study.

To exemplify this procedure, this section presents a complete example applied to robotic welding. The setup (Figure 4.11) is composed by an anthropomorphic robot manipulator (ABB IRB1400 – M2005), the robot controller (ABB IRC5, running version 5.06 of the controller operating system) and a MIG/MAG welding power source (ESAB LUA 315R).

To command and monitor the welding application from the remote computer the following services (implemented by the TCP/IP socket server) must be available at the robot controller:

1. **Robot and robot program control and supervision services:** the remote user must be able to change the robot operational state, start/stop selected robot programs (tasks), and read robot controller and robot programs

actual states. Since actual robot control systems are multitasking systems, users must consider robot programs as tasks and deal with them as such.

2. **IO read/write services:** the remote client needs to monitor and change the actual state of selected IO signals, in a way to safely control the welding application: for example, signals controlling the GAS valve, the WIRE speed, the welding CURRENT, *etc.*, are IO signals generated by the robot controller. Consequently, the possibility to read/write to any digital and analog signal is needed.
3. **Variable read/write services:** if the remote client is necessary to parameterize the welding operation, including the welding trajectory (obtained for example from a CAD software package), to control the welding tasks using commanding variables (see the previous section), to monitor the welding operation, *etc.*, then variable read/write services are needed.

Building a TCP/IP server to implement these types of services is very simple. In the following and with the objective of demonstrating how it could be done, a simple TCP/IP server that explores the facilities available in the new ABB IRC5 robot controller will be designed and built. The choice for this robot controller is based on the fact that it allows users to explore and program TCP/IP socket connections since those are fully supported by the programming environment [43].

The server presented below was written in RAPID (ABB robot programming language [43]), and the client example was coded using the Visual Basic compiler of the Microsoft Visual Studio .NET 2003 developing suite [42]. In the following the software will be designed and built step-by-step.

First step - define the messaging syntax to use: to develop the TCP/IP server to run on the robot controller, a proper messaging syntax should be clearly defined. For this simple example the following message structure will be used:

Command structure:

command parameter_1 parameter_2 ... parameter_n

Answer structure:

error_code answer_1 answer_2 ... answer_n

where "*command*" is a string that identifies the specific command, "*parameter_1*" to "*parameter_n*" are optional parameters associated with the specific command, "*error_code*" is a numeric value that specifies if the command was executed correctly (zero returned) or with errors (negative value identifying the returned error), and "*answer_1*" to "*answer_n*" are the results of the execution of the command. The commands considered are summarized in Table 4.2.

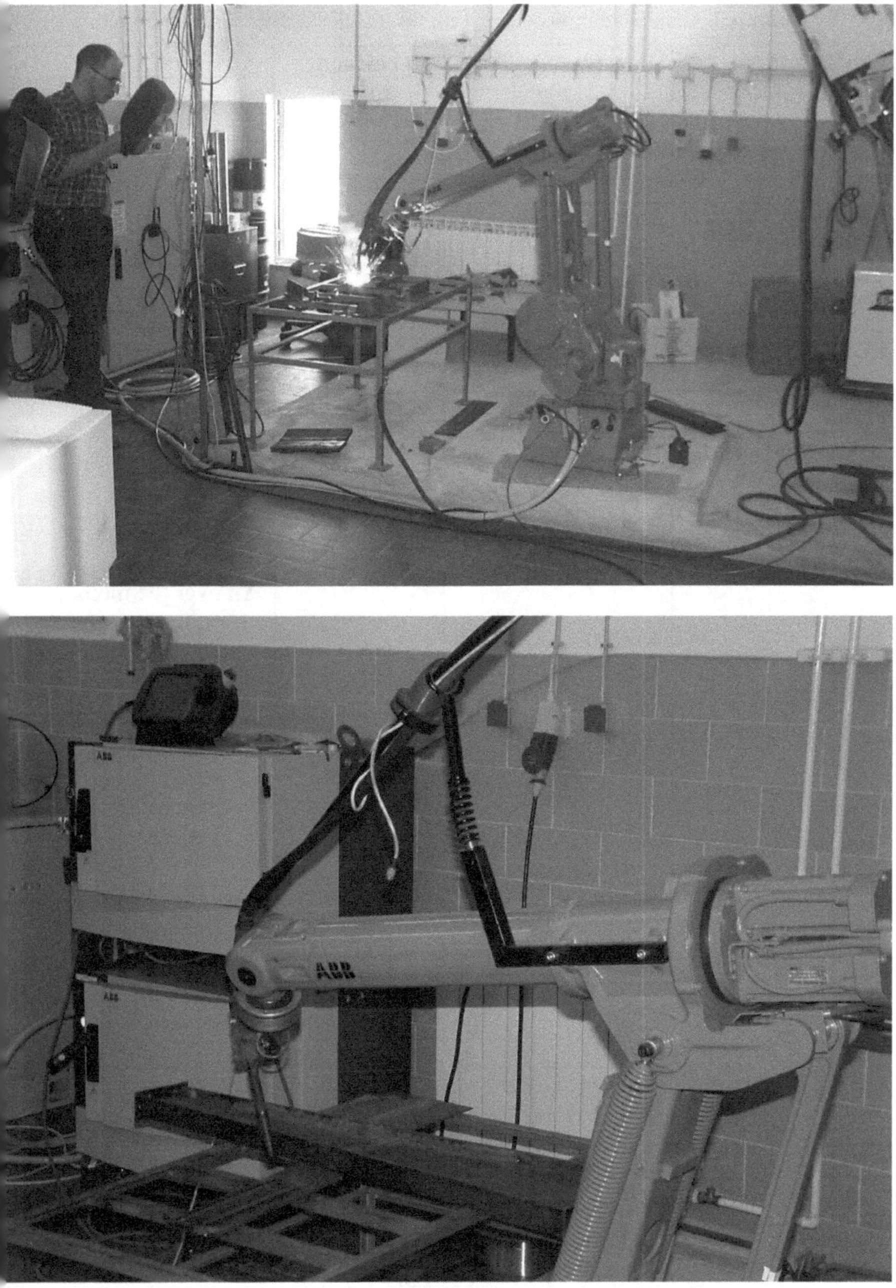

Figure 4.11. Experimental setup used for the TCP/IP sockets server example (ABB IRB1400 robot + ABB IRC5 robot controller)

Table 4.2. Commands considered with the TCP/IP socket server example

Command	Parameters	Error code	Answers
motor_on	-	0 – OK -1 - Failure	-
motor_off	-	0 – OK -1 – Failure	-
program_run	**Par_1:** "name of routine"	0 – OK -1 - Failure	
program_run_pp	-	0 – OK -1 – Failure	
program_stop	-	0 – OK -1 - Failure	
system_state		0 – OK -1 - Failure	**Answer_1:** controller 1 – unknown 2 – stand by 3 – run 4 – emergency **Answer_2:** memory -1 – unknown Value – free memory **Answer_3:** program 1 – unknown 2 – ready 3 – executing 4 – stopped **Answer_4:** operational 1 – unknown 2 – manual 3 – auto 4 – emergency
variable_read	**Par_1:** "name" **Par_2:** type	0 – OK -1 - Failure	**Answer_1:** value
variable_write	**Par_1:** "name" **Par_2:** type **Par_3:** value	0 – OK -1 - Failure	
io_read	**Par_1:** "name" **Par_2:** type	0 – OK -1 - Failure	**Answer_1:** value
io_write	**Par_1:** "name" **Par_2:** type **Par_3:** value	0 – OK -1 - Failure	

Note: if the server returns the error_code = -9999 that means that an illegal or non-existent command was sent.

Considering the specification of commands presented in Table 4.2 the following are examples of commands and possible answers:

Change the robot operational state to "run"
Command: "motor_on"
Answer: 0

Change the state of robot task "main" to "executing"
Command: "program_run" "main"
Answer: 0

Read back the robot system state
Command: "system_state"
Answer: 0 3 98 3 3
(Means: OK, run, 98% memory free, executing, auto)

Read a variable from the robot
Command: "variable_read" "current" "numeric"
Answer: 0 200

Command: "variable_read" "velocity" "numeric"
Answer: 0 10

Command: "variable_read" "home" "position_orientation"
Answer: 0 x y z q1 q2 q3 q4 cf1 cf4 cfx ex1 ex2 ex3 ex4 ex5 ex6
(Means: OK, Cartesian position, quaternion, configuration matrix, external axis) [43],[44]

Write to a variable of the robot
Command: "variable_write" "current" "numeric" 210
Answer: 0

Read a IO signal from the robot
Command: "io_read" "wire_feed" "analog"
Answer: 0 0.35

Command: "io_read" "gas" "digital"
Answer: 0 1

Second step – write the TCP/IP socket server program: as mentioned above this server was implemented using the programming language RAPID to run on an ABB IRC5 robot controller. This fact does not constitute a limitation, since the facilities used here are also available in any multitasking operating system capable of managing and using TCP/IP socket connections. The server code is presented and commented in Figure 4.12.

```
MODULE server_sock
VAR socketdev server_socket;
VAR socketdev client_socket1;
VAR string receive_string;
VAR string command;
VAR string answer_string;
VAR string par_1;
VAR string par_2;
VAR num par_3;
VAR num answer_1;
VAR num answer-2;
VAR num answer_3;
VAR num answer_4;
VAR string client_ip;
VAR socketstatus state;
PERS num retry_sock:=0;
PERS num working:=1;
```

Declaration of variables

```
PROC main()
  WaitTime 0.1;
  SocketCreate server_socket;
  SocketBind server_socket, "172.16.0.89", 2004;
  SocketListen server_socket;
  WHILE TRUE DO
```

Create and bind the server socket and start listening for connections

```
again_sock:
    IF working = 1 THEN
      SocketAccept server_socket, client_socket1 \ClientAddress:=client_ip;
      IF retry_sock = 1 THEN
        retry_sock:=0;
        GOTO again_sock;
      ENDIF
      SocketReceive client_socket1 \Str := receive_string;
      extract_info;
      TEST command
        CASE "motor_on":
          motor_on; SocketSend client_socket1 \Str := answer_string;
        CASE "motor_off":
          motor_off; SocketSend client_socket1 \Str := answer_string;
        CASE "program_run":
          program_run(par_1); SocketSend client_socket1 \Str :=
          answer_string;
        CASE "program_run_pp":
          program_run_pp; SocketSend client_socket1 \Str := answer_string;
        CASE "program_stop":
          program_stop; SocketSend client_socket1 \Str := answer_string;
        CASE "program_run_pp":
          program_run_pp; SocketSend client_socket1 \Str := answer_string;
```

State machin

```
            CASE "system_state":
                system_state; SocketSend client_socket1 \Str := answer_string;
            CASE "variable_read":
                variable_read(par_1, par_2); SocketSend client_socket1 \Str :=
                answer_string;
            CASE "variable_write":
                variable_write(par_1, par_2, par_3); SocketSend client_socket1
                \Str := answer_string;
            CASE "io_read":
                io_read(par_1, par_2); SocketSend client_socket1 \Str :=
                answer_string;
            CASE "io_write":
                io_read(par_1, par_2, par_3); SocketSend client_socket1 \Str :=
                answer_string;
            DEFAULT:
                SocketSend client_socket1 \Str := -9999;
        ENDTEST
        SocketClose client_socket1;
      ELSE                                    --> Ignore connections if server
        WaitTime 0.1;                             is disabled
        SocketClose client_socket1;
      ENDIF
    ENDWHILE
    ERROR                                     --> Retry socket listen if
    SocketClose client_socket1;                   socket timed out
    retry_sock:=1;
    TRYNEXT;
    UNDO
    SocketClose server_socket;
    SocketClose client_socket1;
ENDPROC
ENDMODULE
```

Figure 4.12. TCP/IP socket server RAPID code [43][44]

This server was placed on the controller running as an extra-task, *i.e.*, the robot has a main task, where the foreground applications are running, and a second task used to communicate with the external computer: to receive commands and exchange information.

Third step – write the client program: any TCP/IP socket server should be capable of receiving and accepting connections from client applications, receive and process commands, and deliver the obtained results. The following example (Figure 4.13) is a simple TCP/IP socket client program coded using the Microsoft Visual Basic .NET 2003 developing suite.

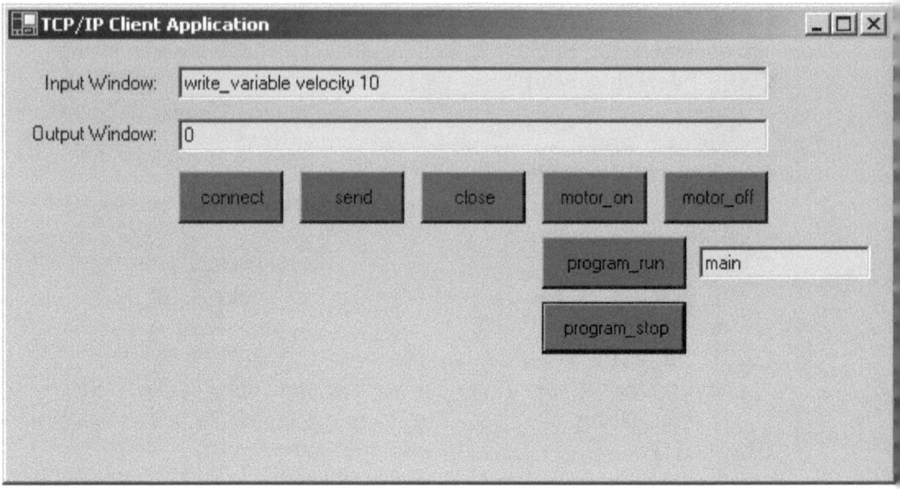

Figure 4.13. TCP/IP socket client.

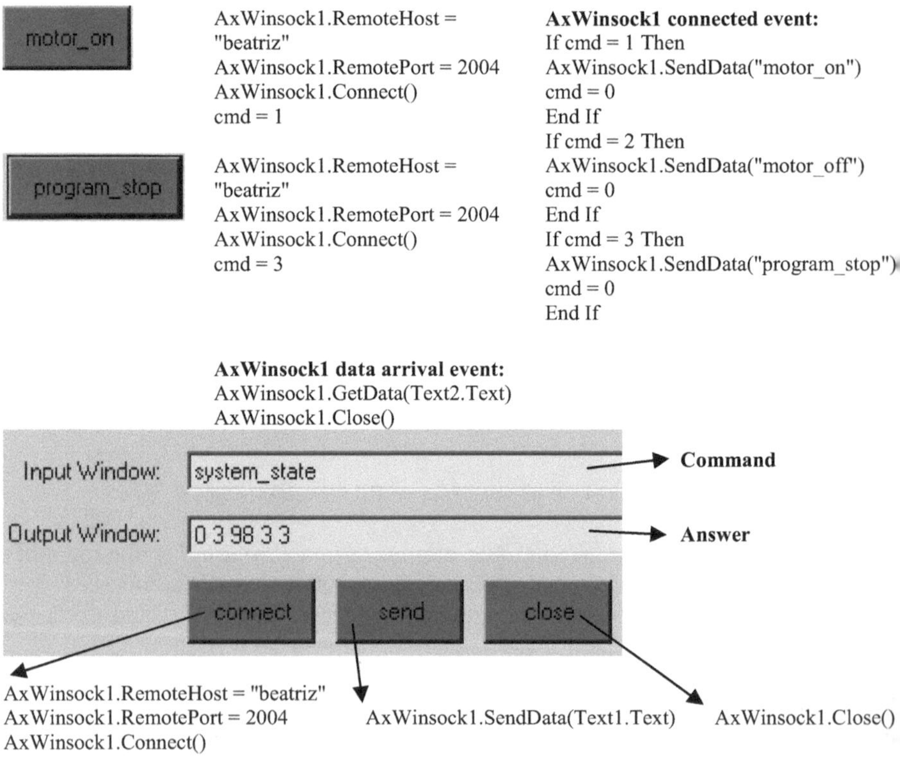

Figure 4.14. Code detail for the client software

The client program implements calls to the TCP/IP server program defined above, using the Microsoft Winsock2 implementation, and was written to be a very simple example. A few software buttons were added to the client program to perform certain operations, but with this program the user can also write the full instruction in the command window, send it to the robot and receive the answer in the output window. To demonstrate the effort involved in sending commands and receiving answers, the following (Figure 4.14) shows the code associated with some of the software buttons.

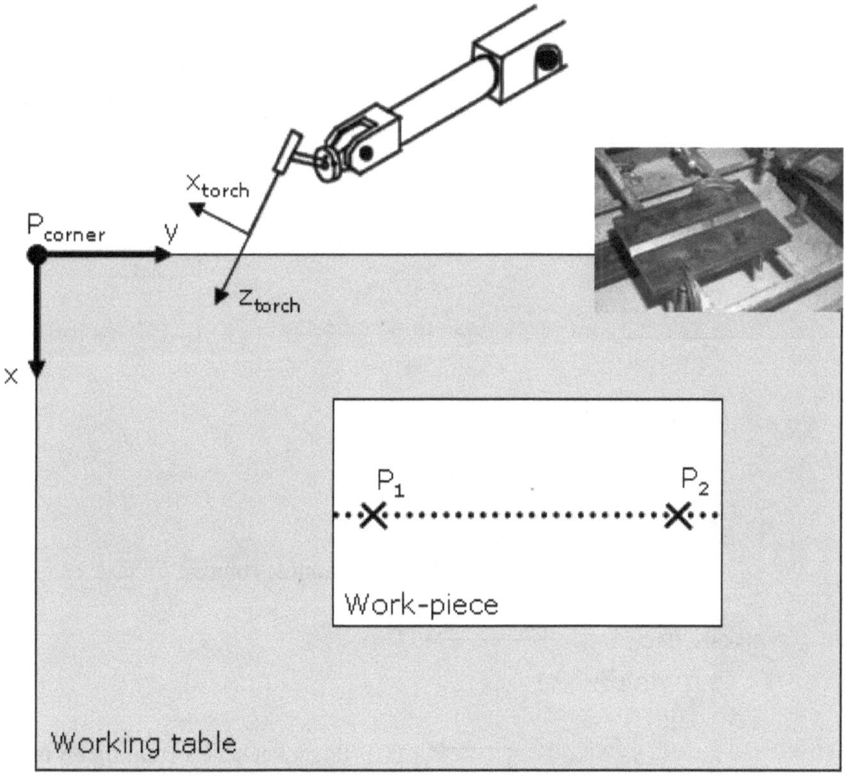

Figure 4.15. Simple welding application used for demonstration

4.6 Simple Welding Example

Let's consider a simple welding application (Figure 4.15), where the robotic welding system (Figure 4.11) is commanded to execute a linear welding trajectory between two positions: P_1 and P_2. The user should be able to acquire and adjust the two mentioned positions, but also to change the welding parameters: *voltage*, *current* and *welding speed*. From the two technologies presented above to implement robot controller access, the example presented in this section will use the TCP/IP socket technology since the RPC technology is alternatively used in Chapter 5. The simple TCP/IP socket server presented in Figure 4.12 was designed to operate in a multitasking environment (Figure 4.16), where the server runs in one task (task 2) and the operational routines related with the specific welding application run in another task (task 1). That's the case of the robot controller used with this example, (ABB IRC5) which is running the *ABB RobotWare* operating system (version 5.06) [43] with the option multitasking activated.

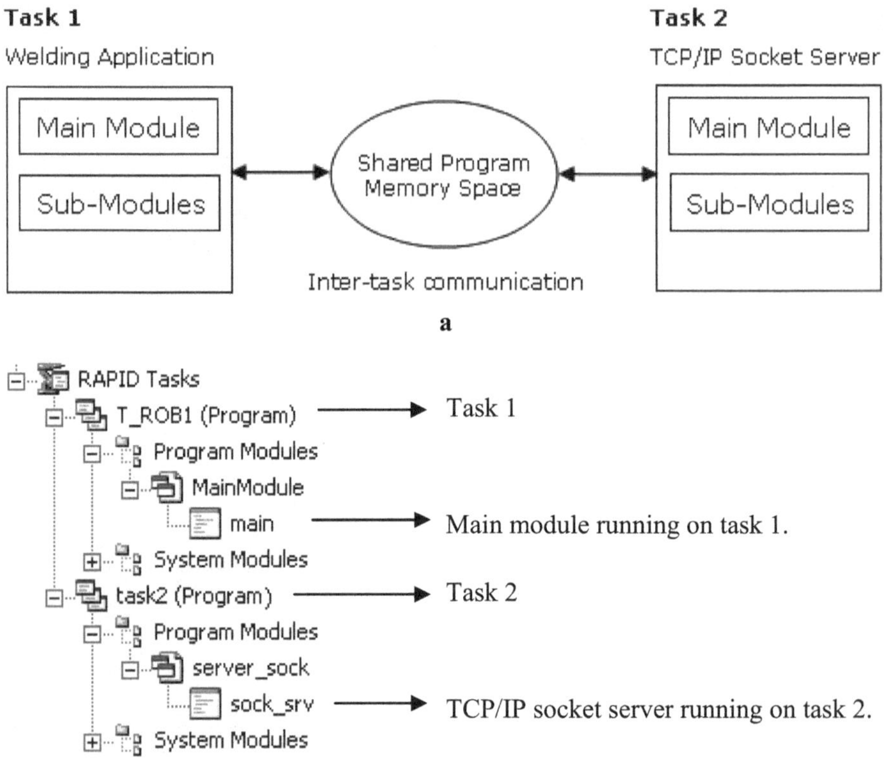

b

Figure 4.16 a, b. Multitasking environment: **a** - diagram showing how tasks communicate using shared memory space; **b** - aspect of the *RobotStudio* [43] RAPID tasks view, on the PC side, showing the running tasks

Consequently, the goal here is to design the welding application task in such a way as to offer the specific functionalities needed to be able to run the planned welding operations. Considering that those functionalities are to be requested remotely, it seems logical to implement a basic SWITCH-CASE-DO loop driven by a variable controlled from the remote computer (see Section 4.5.1). It should be noted that the remote client can change the value of any program variable just by using the "*write_variable*" service implemented in the TCP/IP socket server (Figure 4.12).

Therefore the following services should be implemented in task 1:

1. Move the robot in Cartesian space: since the piece to weld is placed on a welding table (Figure 4.15), it's enough to allow the user to move the robot relative to a fixed position (P_{corner}) on the table. To move the robot parallel to the surface, avoiding collisions, the user must be able to send jogging commands to the robot in the form of Cartesian offsets and rotation offsets about the axis of the defined robot TOOL (Figure 4.15). To add the jogging capability the position P_{corner} should be acquired to the system aligning the TOOL reference frame with the defined position frame. The position obtained from P_{corner} adding 200mm in the Z direction (WORLD reference frame) can be called "home" or "safe position",

$$P_{home} = \text{Offsets}(P_{corner}, 0, 0, 200);$$

where the function *Offsets* [43] adds 200mm to the Z Cartesian direction, recalculating the orientation parameters to keep the orientation in the new Cartesian position.

Therefore, to jog safely the robot in the welding working space, the following two services must be available from the welding application:

Service "Go Home"
CASE 1000:

```
  MOVEJ Offsets(Pcorner , 0, 0, 200), v100, fine, tool_torch;
decision1:=0;
```

Service "Move_Cartesian_Linear"
CASE 1001:

```
WHILE decision1 = 1001 DO
 IF move <> 0 THEN
   MOVEJ RelTool(Pcorner , x, y, z\RX:=rx, RY:=ry, RZ:=rz), v100, fine, tool_torch;
 ENDIF
ENDWHILE
decision1:=0;
```

An extra service capable of moving the robot in coordinated joint motion could be interesting to solve motion difficulties related with singular zones. The service could be obtained just by adapting the previous "Move_Cartesian_Linear" service to:

Service "Move_Cartesian_Linear_Joint"
CASE 1001:

```
Pactual:= CRobT(\Tool:=tool_torch);
WHILE decision1 = 1001 DO
```

```
IF move <> 0 THEN
 IF move_type = 1 THEN
  MOVEL RelTool(P_actual ,x, y, z\RX:=rx, RY:=ry, RZ:=rz), v100, fine, tool_torch;
 ELSE
  MOVEJ RelTool(P_actual , x, y, z\RX:=rx, RY:=ry, RZ:=rz), v100, fine, tool_torch;
 ENDIF
ENDIF
ENDWHILE
decision1:=0;
```

Linear Motion (MOVEL)

Joint Coordinated Motion (MOVEJ)

2. Acquire positions P1 and P2: this is a very simple task done by the following services:

Acquire position P_1
CASE 1002:

```
P1:=CRobT(\Tool:=tool_torch);
decision1:=0;
```

Acquire position P_2
CASE 1003:

```
P2:=CRobT(\Tool:=tool_torch);
decision1:=0;
```

3. Start and stop weld: these are services used to initiate the arc doing the weld, and to stop the arc:

CASE 1004:

```
weld_on;
decision1:=0;
```

CASE 1005:

```
weld_off;
decision1:=0;
```

The procedures *weld_on* and *weld_off* are presented in (Figure 4.17), along with sub-procedures used by them. The *weld_on* procedure implements the timing defined by the welding power source, represented in Figure 5.2.

4. Approach position P1: this service is used to move the robot safely to P1:

CASE 1006:

```
MOVEJ Offset(P1, 0, 0, 200), v200, z50, tool_torch;
MOVEJ Offset P1, v100, z50, tool_torch;
decision1:=0;
```

```
PROC weld_on()
 status:=0;
 CONNECT timeint WITH HEAT_Timer;
 set_ignition;
 SetDO doGAS,1;
 IF ptime<>0 THEN
  WaitTime ptime;
 ENDIF
 SetDO doFEED,1;
 SetDO doWELD,1;
 WaitUntil diARC_EST=1\MaxTime:=1\TimeFlag:=oops;
 IF oops=TRUE THEN
  werro:=801;
  SetDO doGAS,0;
  SetDO doFEED,0;
  SetDO doWELD,0;
  reset_signals;
  EXIT;
 ENDIF
 IF imtime<>0 THEN
  WaitTime imtime;
 ENDIF
 set_heating;
 ITimer\Single,htime,timeint;
ENDPROC

PROC set_ignition()
 SetAO aoWD_REF,wd_iref;
 SetAO aoFEED_REF,feed_iref;
 status:=1;
ENDPROC
PROC set_heating()
 SetAO aoWD_REF,wd_href;
 SetAO aoFEED_REF,feed_href;
 status:=2;
ENDPROC

PROC set_references()
 SetAO aoWD_REF,wd_ref;
 SetAO aoFEED_REF,feed_ref;
 status:=3;
ENDPROC

PROC reset_signals()
 SetDO doGAS,0;
 SetDO doFEED,0;
```

Heating Timer (CONNECT timeint WITH HEAT_Timer;)

Set the references for the ignition period (set_ignition;)

Start wire feed and establish the arc (SetDO doFEED,1;)

Check if the arc was established and reset signals if it wasn't (WaitUntil diARC_EST=1\MaxTime:=1\TimeFlag:=oops;)

Set the references for the heating period (set_heating;)

```
  SetDO doWELD,0;
  SetAO aoWD_REF,0;
  SetAO aoFEED_REF,0;
  status:=-1;
 ENDPROC

 PROC weld_off()  ------------>        Stop welding pocedure
  status:=4;
  SetDO doFEED,0;
  WaitTime btime;
  SetDO doWELD,0;
  WaitTime ctime;
  SetDO doFEED,1;  ------------>       Extra-welding pulse
  SetDO doWELD,1;
  WaitTime ftime;
  SetDO doFEED,0;
  WaitTime btime;
  SetDO doWELD,0;
  WaitTime ctime;
  SetDO doGAS,0;
  reset_signals;                        Set the welding references
 ENDPROC                                after heating period
                                        (executed on the heating
 TRAP HEAT_Timer  ------------>         timer service routine)
  set_references;
  IDelete timeint;
 ENDTRAP
```

Figure 4.17. Starting and stopping welding

5. Execute a linear weld from P1 to P2: this service executes the planned weld:

```
CASE 1007:
weld_on
MOVEL P2, velocity, fine, tool_torch;
weld_off;
MOVEJ Offset(P2, 0, 0, 200), v100, z50, tool_torch;
MOVEJ Offset(Pcorner, 0, 0, 200), v200, z50, tool_torch;
decision1:=0;
```

Having these procedures implemented, it is fairly simple to command the planned welding operation, along with the acquisition of the two welding points P1 and P2, the robot jogging operation, *etc.*, from a remote computer. A special tool may be built for the case, but the code already used for the simple TCP/IP socket client (Figure 4.13 and Figure 4.14) can be used for a demonstration. In that case, the following simple commands can be sent to access the above services (see also Table 4.2):

Send the robot to position "Home"
Connect
Send message: variable_write decision1 num 1000
Close

Move the robot 25mm in the positive X direction and 30mm in the positive Y direction

Write 25 to X
Connect
Send message: variable_write x num 25
Close

Write 30 to Y
Connect
Send message: variable_write y num 30
Close

Select Linear Motion
Connect
Send message: variable_write move_type num 1
Close

Call the Service "Move_Cartesian_Linear_Joint"
Connect
Send message: variable_write decision1 num 1001
Close

Move the robot now
Connect
Send message: variable_write move num 1
Close

Other motions can be commanded now writing to x, y, z, rx, ry, rz, and setting move = 1 after that to effectively move the robot.

To leave the service
Connect
Send message: variable_write decision1 num 0
Close

Acquire P1
Connect
Send message: variable_write decision1 num 1002
Close

Acquire P2
Connect
Send message: variable_write decision1 num 1003
Close

Set the welding current
Connect
Send message: variable_write feed_ref num value
Close

Note: the *welding current* regulates the *wire feed* reference signal (*feed_ref*) and is commanded using a converting chart given by the manufacturer of the welding power source.

Set the welding voltage
Connect
Send message: variable_write wd_ref num value
Close

Note: the *welding voltage* regulates the *wd_ref* signal and is commanded using a converting chart given by the manufacturer of the welding power source.

Execute the weld from P1 to P2
Connect
Send message: variable_write decision1 num 1007
Close

Note: before sending this command the user must set all the timings used in the welding routines presented in Figure 4.17.

Read actual value of variable *decision1*
Connect
Send message: variable_read decision1 num
Close

It's straightforward to associate these procedures to software *buttons* and build a nice client software tool to command easily the presented welding example (hiding the code from the user). This process can be done using Visual Basic or Visual C++, or any other programming language, providing that the Microsoft Winsock2 software API, or any other equivalent API, is included in a way to add socket management and communication capabilities to the project.

4.7 Semi-autonomous Manufacturing Systems

Actual manufacturing systems need to have high degrees of autonomy requiring less operator intervention to improve agility and efficiency. In fact, if the manufacturing systems are designed to operate remotely using simple commands and parameterization, then the tasks necessary to adjust the setup to manufacture a different product model may be fully automatic. This procedure should lead to big improvements in terms of productivity, along with considerable gains in terms of agility and flexibility. In the particular case of robotic welding, the following are examples of things that can be done without operator intervention:

1. Change the welding parameters: current, voltage and speed.
2. Change the way the welding power source is controlled (Figure 5.2).
3. Change the welding sequence just by adjusting the welding points, the trajectories between them and the definition on what are the welding and the approach/escape trajectories.

This can easily be implemented by properly designing the services offered by the robot controller application software, having them working based on the remote parameterization. Furthermore, to include the robotic welding cell in a manufacturing line, an "automatic mode" service must be designed with the objective of having the system responding to sensor signals and to PLC commanding signals. In fact, depending on the application, a few sensors are needed to inform about the necessary conditions to execute the task, which includes information like: "work-piece in place", "power source on-line", "welding gun clean", "cell not violated", *etc.* It is also very important to interface with the PLC managing the manufacturing line, because the coordinating signals that make the manufacturing sequence occur are generated by the software running on the PLC (usually specified using GRAFCET).

To demonstrate how this can be done, let's consider the simple welding process presented in Section 4.6. The services discussed there are to be used in "local or manual mode" to adjust points, simulate the procedure, *etc.* They shouldn't be used in "automatic mode", where the robot needs to react to sensor information and to coordinating signals coming from the PLC managing the manufacturing line. Figure 4.18 represents a possible manufacturing line where the simple welding example was inserted. The system is composed by a conveyor belt that carries the pieces to weld, a centering and holding pneumatic device that holds the pieces during the welding procedure and a collection of position sensors, responsible for detecting the piece to stop the conveyor and to verify if the work-piece is in position after being trapped by the pneumatic cylinder. Only after the previous tasks are completed is the robot commanded to perform the welding operation. When the operation is done, the robot should signal the controlling PLC to free the welded piece, re-start the conveyor motion and wait for another one to weld.

Consequently, the "automatic mode" cycle should look like the following:

```
CASE 9000:
  WHILE decision1 = 9000 DO
    WaitUntil di_piece_in_place = 1;
    SetDO piece_weld_done, 0;

    ! Approach P1
    MOVEJ Offset(P1, 0, 0, 200), v200, z50, tool_torch;
    MOVEJ Offset P1, v100, z50, tool_torch;

    ! Weld from P1 to P2
    weld_on
    MOVEL P2, velocity, fine, tool_torch;
    weld_off;

    ! Move to SAFE position
    MOVEJ Offset(P2, 0, 0, 200), v100, z50, tool_torch;
    MOVEJ Offset(Pcorner, 0, 0, 200), v200, z50, tool_torch;

    ! Signal handshake: end of cycle
    SetDO piece_weld_done, 1;
    WaitUntil di_piece_in_place = 0;
  ENDWHILE:
  decision1:=0;
```

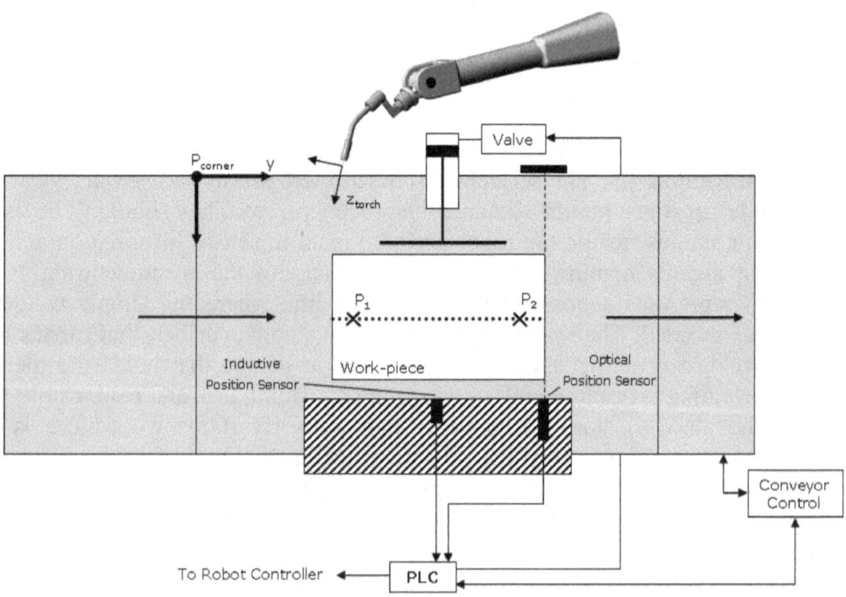

Figure 4.18. Simply welding example include in a manufacturing line.

To operate the system, using the same TCP/IP sockets strategy, the calling computer should send the welding parameterization, using the available services, and command "automatic mode" to start production. This can be done from any computer on the network, namely from computers running the software used to track and control the production. In many industries, production is closed tracked in any part of the manufacturing cycle, which is composed by several in-line manufacturing systems that perform the necessary operations transforming the raw materials in a final product. In many cases, if properly designed, those individual manufacturing systems require simple parameterization (like the robot welding tasks) to execute the tasks they are designed to execute. If that parameterization can be commanded remotely by automatic means from where it is available, then the system becomes almost autonomous in the sense that operator intervention is reduced to the minimum and essentially related with error and maintenance situations. A system like this will improve efficiency and agility, since it is less dependent on human operators. Also, since those systems are built under distributed frameworks, based on client-server software architectures which require a collection of functions that implement the system functionality, it is easier to change production by adjusting parameterization (a software task now) which also contributes to agility. In adition, since all information about each item produced is available in the manufacturing tracking software, it is logical to use it to command some of the shop-floor manufacturing systems, namely the ones that require simple parameterization to work properly. This procedure would take advantage of the available information and computing infrastructure, avoiding unnecessary operator interfaces to command the system. Also, further potential gains in terms of flexibility and productivity are evident.

Furthermore, to adjust the welding parameters and start production in the simple welding example presented, the collection of commands would be:

Set the welding current
Connect
Send message: variable_write feed_ref num value
Close

Set the welding voltage
Connect
Send message: variable_write wd_ref num value
Close

Set the welding velocity
Connect
Send message: variable_write velocity num value
Close

Note: these three commands are used to define the welding parameters. They can be done once, and/or adjusted during production, or done asynchronously in any combination, without being followed by the "automatic mode" command. This command requires a previous definition of the welding parameters, because it executes welding trajectories, but since the values of those parameters are kept there's no need to send them again between two consecutive production cycles if they aren't to be changed.

Command "automatic mode"
Connect
Send message: variable_write decision1 num 9000
Close

Finally, it should be emphasized that the welding parameters can be adjusted without restrictions at any time during "automatic mode" (a service implemented in task 1), since those are calls to the TCP/IP services running in a different task (task 2).

4.8 Chapter Final Notes

In this chapter several system issues related to robotic welding were discussed with the triple objective to:

1. **Clarify:** robotic welding is very demanding on system resources and capabilities mainly because automatic welding is a task that requires an adapting behavior of the automatic system [19],[23]. This has an enormous impact on the requirements for the system planned to make the welding task automatic. First because the system needs to handle the welding process models (databases and/or knowledge base [45]) to enable automatic selection of the welding parameters and their dynamic behavior based on the observed welding conditions. Second because the automatic system needs to accommodate the advanced sensors used to obtain that information, but also to perform seam tracking, inspection and quality analysis. The required sub-systems and software modules necessary to incorporate information from sensors must be flexible enough to allow multiple sensors to be integrated easily and be fully explored.
2. **Contribute:** to cope with the robotic welding requirements there are several features from the robot controllers that can be used, along with available software architectures, software component technologies, communication data and signal protocols, sensors and interfaces, *etc.* The choice is up to the system engineers to select from the available possibilities, although a few standard approaches are obvious candidates as this chapter clearly pointed out. Any adopted solution should use a software framework capable of handling local and remote services with the objective of incorporating the necessary functionalities using currently available industrial standards.
3. **Demonstrate:** in this chapter a software architecture was defined (Figure 4.9) and several examples on using it were given with the objective of

clarifying the issues described, but also to show how they could be implemented using actual equipment and software tools.

This chapter also discusses the construction of manufacturing systems that incorporate high degrees of autonomy with the objective of improving productivity and efficiency of the manufacturing processes. That means distributing functions to all the components of the system on a client-server framework. Furthermore, the software must be designed as general as possible to offer the system functionality to the remote clients in the form of customable services. Furthermore, for the remote client any of the system features can be parameterized and in this way tailored to the user needs, which means adjusted to execute the production and operational changes required for the actual installation. This means network interfaces, based on the TCP/IP protocol, and remote procedure calls, enabling direct command of shop-floor manufacturing setups from anywhere in the factory.

The following chapter presents and discusses a complete robotic welding system. The chapter is very technical and goes deeply into implementation details with the objective of being informative and demonstrative. The chapter also demonstrates some of the features discussed in this book using a particular robot/controller and the software architecture presented in this chapter.

4.9 References

[1] Pires, JN, and Loureiro, A *et al*, “Welding Robots”, IEEE Robotics and Automation Magazine, June, 2003

[2] Pires, JN, and Loureiro, Altino *et al*, “Object-Oriented and Distributed Software applied to Industrial Robotic Welding”, Industrial Robot, An International Journal, MCB University Press, March 2002.

[3] Olsson M, “Simulation and execution of autonomous robot systems”, Ph.D. thesis, Devision of Robotics, Department of Mechanical University, Lund Institute of Technology, Lund University, Sweden, 2002.

[4] Yu, JY, Na, SJ, “A Study on Vision Sensors for Seam Tracking of Height-Varying Weldment. Part1: Mathematical Model”, Elsevier Mechatronics Journal, Vol, 7, Nº 7, pp. 599-612, 1997.

[5] Yu, JY, Na, SJ, “A Study on Vision Sensors for Seam Tracking of Height-Varying Weldment. Part2: Applications”, Elsevier Mechatronics Journal, Vol, 8, Nº 1, pp. 21-36, 1998.

[6] Bolmsjo G., “Sensor System in Arc Welding”, Technical Report, Lund Institute of Technology, Production and Materials Engineering Department, 1997.

[7] Agren B. "Sensor Integration for Robotic Arc Welding." PhD thesis, Lund University, 1995

[8] Smartt HB *et al*, “Intelligent Control of Modeling Welding Cell”, Proceedings of the ACM 6th International Conference on Trends in Welding Research, Pine Mountain, USA, April 2002.

[9] Bolmsjo G, et al, “Off-line programming of GMAW robotic systems – a case study”, International Journal for the Joining of Materials, Vol.9, Nº3, pp.86-93, 1997.

[10] Fridenfalk M, "Development of intelligent robot systems based on sensor control" Ph.D. thesis, Division of Robotics, Department of Mechanical University, Lund Institute of Technology, Lund University, Sweden, 2003.
[11] Ericsson M, "Simulation of robotic TIG-Welding" Department of Technology University of Trollhättan (Uddevalla), Trollhättan, Sweden, 2003.
[12] Cook, GE, "Robotic Arc Welding: Research in Robotic Feedback Control", IEEE Transactions on Industrial Electronics, Vol. 30, N° 3, August 2003.
[13] Loureiro A., Velindro M. and Neves F., "The influence of Heat Input and the Torch Weaving Movement on Robotized MIG Weld Shape", International Journal for the Joining of Material, Vol.10, N.3/4, pp. 86-91,1998.
[14] Bolmsjö G., Olsson M., Nikoleris G., and Brink K. Task programming of welding robots. In Proceedings of the Int. Conf. on the Joining of Materials, JOM-7, pages 573-585, May-June 1995.
[15] Richardson RW, "Robotic Weld Joint Tracking Systems - Theory and Implementation Methods", Welding Journal, Nov. 1986.
[16] Books B., "Welding Robots - The State of the Art", Welding and Metal Fabrication June 1991.
[17] Drews P., Starke G., "Development Approaches for Advanced Adaptive Control in Automated Arc Welding", Internal Report XII-970, Mechatronics Deartment, Achen, Germany, 1986.
[18] Yada Y. *et al*, "Development of Real Time Welding Control System by Using Image Processing", Internal Report XII-998, Mechatronics Department, Achen, Germany, 1986.
[19] Proceedings of the International Workshop on Robotic Welding Systems and Process Monitoring (http://robotics.dem.uc.pt/welding/), Coimbra Portugal, January 2001, published at the Journal Robótica, January 2001.
[20] Boillot, JP *et al*, "The Benefits of Vision in Robotic Arc Welding", ServoRobot Inc., http://www.servorobot.com, February, 2002.
[21] Sicard, P, Levine, MD, "An Approach to an Expert Robot Welding System", IEEE Transactions on Systems, Man and Cybernetics, Vol. 18, N° 2, March/April, 1988.
[22] Miller, M *et al*, "Development of automated real-time data acquisition system for robotic weld quality monitoring", Elsevier Mechatronics, Vol 12, pp. 1259-1269, 2002.
[23] Bolmsjö G., "Knowledge Based Systems in Robotized Arc Welding. Knowledge Based Systems - Advanced Concepts", Techniques and Applications, (Spyros Tzafestas, editor). Chapter 17, pp. 465-495. World Scientific, ISBN 9810228309, 1997.
[24] Magy E, Haidegger G, Object-Oriented Approach for Analyzing, Designing and Controlling Manufacturing Cells, Proceedings of AUTOFACT'93, Chicago, 1993.
[25] Li, L, Object-Oriented Modeling and Implementation of Control Software for a Robotic Flexible Manufacturing Cell, Robotics and Computer Integrated Manufacturing, 1994.
[26] Dahlén P. and Bolmsjö G., "Human Factors in the Justification of Advanced Manufacturing Systems". International Journal of Human Factors in Manufacturing, 6(2):147-162, 1996.
[27] Fahim, Choi K, "The UNISET approach for the Programming of Flexible Manufacturing Cells", Robotics and Computer Integrated Manufacturing, 1998.
[28] OAEnterprise, Object Automation Inc (http://www.oa.com), 1999.
[29] Osaca Arquictecture, Osaca organization, (http://www.osaca.org), 2001.
[30] Factory Suite A^2, WonderWare Inc. (http://www.wonderware.com), 2004
[31] Iwanitz F, Lange J, "OPC - Fundamentals, Implementation and Application", Hüthig Fachverlag Editora and OPC-Foundation, 2005
[32] Bloomer J., "Power Programming with RPC", O'Reilly & Associates, Inc., 1992.

[33] Cover D, Stevens D, "Internetworking with TCP/IP", Vol I to III, Prentice Hall, 2nd Edition, 1998.
[34] Halsall F., "Data Communications, Computer Networks and Open Systems", Third Edition, Addison-Wesley, 1992.
[35] Rogerson D., "Inside COM", Microsoft Press, 1997.
[36] Box D., "Essential COM", Addison-Wesley, 1998
[37] "Introduction to OSF DCE, OSF DCE", Open Software Foundation and Prentice Hall, 1995.
[38] RAP, Service Protocol Definition, ABB Flexible Automation, 1996 - 2004.
[39] Pires JN, Sá da Costa JMG, "Object Oriented and Distributed Approach for Programming Robotic Manufacturing Cells", IFAC Journal on Robotics and Computer Integrated Manufacturing, February 2000.
[40] Sá da Costa, JMG, Pires, JN, "Future Welding Robot Developments", Journal Robótica, nº41, January 2001.
[41] Pires, JN, "Complete Robotic Inspection Line using PC based Control, Supervision and Parameterization Software", Elsevier and IFAC Journal Robotics and Computer Integrated Manufacturing, Vol. 21, Nº1, 2005
[42] Microsoft Studio .NET 2003, TechNet On-line Documentation, Microsoft Corporation, http://www.microsoft.com, 2003.
[43] ABB IRC5 Documentation, ABB Flexible Automation, 2005.
[44] ABB IRB1400 Users Manual, ABB Flexible Automation, 2005.
[45] The Welding Institute, "Standard Data for Arc Welding", Abington Hall, Abington, UK.

5

Robotic Welding: Application Examples

5.1 Introduction

When robots are used on the shop floor to perform welding tasks, users expect massive improvements in terms of flexibility, productivity and quality. But that outcome is not an easy task, since a considerable amount of additional setup and programming work is necessary to guarantee that robots perform as expected with the required quality. This process is time consuming and requires skilled personnel, which means that robots take longer to be programmed and set up than really doing some interesting work, introducing an overhead cost on time and personnel that must be considered. And that is a major problem, since actual concepts like availability and agility [1],[2] are key issues of modern manufacturing. In fact, those concepts are fundamental to have robot automation take a more important part on the shop-floor, cooperating with humans, which is decisive to help modern country companies fight the low salary trap. Machines that store high levels of flexibility, like robot manipulators, but fail in terms of agility because automation integrators and machine builders were unable, when designing a specific machine, to expose that flexibility to the users may become less interesting. Also, for SMEs robot automation becomes uninteresting since all these companies work with small series, under contract by customers, which means several models of products. Knowing that the big majority of the production workforce is represented by SMEs, this scenario constitutes a big problem. Consequently, solving these issues, reducing setup and programming times, is one of the major problems that robot manufacturers must address clearly in the near future.

The welding quality is closely related to the weld bead shape which is regulated by the selected welding parameters (see Chapter 2). Consequently, several databases have been developed, such as those from the Welding Institute – UK [3], to help in selecting the parameters more suitable for the particular situation under consideration. In these databases the input data is generally the type of weld (butt weld or fillet weld), the welding position (flat, horizontal, vertical or overhead), wire diameter and the plate thickness or eventually the leg length in the case of fillet welds. The output data is usually the welding parameters: current, voltage, welding speed and number of weld beads/layers. With databases of this type in the computer the selection of the welding parameters may be done automatically. Even the selection of the wire diameter may be carried out automatically as a function of the thickness of the components or stay for free selection being an input parameter. It would be expectable that, with this information in the computer having a CAD model of the component to be welded, the system could be able to select the welding data for each weld and send it to the robotic welding system. Although it seems easy to achieve this goal in the case of single welds, some data is missing in the available databases for the case of welds with multiple layers. In fact, in this case the position of the torch for each layer needs to be indicated to the robot.

Since that for the majority of the companies that produce multi-layer welds there is only a small number of distinct welds, then it is not hard to fill up the database for their particular case. Consequently, using this method it is easy to carry out the off-line programming of the components to be welded, being only necessary to adjust the coordinates of the process points in the first specimen to be welded.

Furthermore, when robots are used to weld, the programming problem arises immediately. In this chapter a robotic welding example having CAD programming capabilities will be introduced, giving special attention to the implementation details. The subject will be addressed by extending the concepts presented in Section 4.5 along with a simple CAD interface that enable simpler and faster programming. A few test cases will be presented, explained and discussed with the objective of demonstrating that the basic features are available and can be used to build powerful robotic welding solutions.

5.2 A Robotic Welding System

Robots manipulators are essentially position-controlled devices that can receive a trajectory and run it continuously. With welding applications it is necessary to start from a trajectory, given for example from a CAD model of the work-piece, and to have the means to correct it in real time, as function of the observed results of the welding process.

5.2.1 Overview of the System

To achieve the above-mentioned goal, systems for guidance and inspection are needed (see Chapter 3), but also the possibility to correct in real time the position of the robot and the welding parameters, and a computational platform suitable for developing the software necessary to handle the monitoring and control tasks. Unfortunately these features aren't usually available for the following reasons:

1. Actual robot manipulators have closed controllers, not allowing real time position correction by costumer programming.
2. Most of the actual robot controllers do not allow remote control from an external computer.
3. It is very difficult to attach guidance sensors with good performance, because most of the robot controllers are not prepared to do it, or restrict that possibility, not providing fast interfaces freely accessible by advanced users.
4. Robot programming environments are not powerful enough to handle tasks that require complex control techniques (learning, supervisory, adaptive, etc).

With the system introduced here several of these limitations are reduced as follows:

1. A robot control system that allows position correction commanded by a remote computer is used. That's not a standard feature, but was added to the system by using other features available from the controller.
2. A distributed and client-server based software architecture, that enables remote control using Ethernet networks was developed.
3. The necessary sensors may be attached to the computer that controls the robot and not to the robot controller itself. This excludes the seam tracking sensor which requires a low-level interface with the robot controller.
4. A personal computer is used as programming environment, taking advantage of the huge amount of programming and analysis tools available for those platforms.

The basic setup is composed of an industrial robot (model ABB IRB1400 M98), a robot controller (model ABB S4C+ M98), a MIG/MAG welding power source (ESAB LUA 315R) and a computer running Windows XP (any other DCOM [4] based version of the *Microsoft Windows* operating system could have been used). The welding power source is connected to the robot controller IO subsystem, being in this way completely controlled by the robot controller. The robot controller is equipped with an Ethernet board being accessible from the local area network (Figure 5.1) by any computer also connected to the network.

Achieving automatic parameter selection also means using sensing devices not only for guiding the welding torch (joint tracking), but also for real-time acquisition of the welding geometry. Simple torch guidance and joint tracking can be obtained using mechanical, electrical or optical sensors: current sensors (to sense the arc current during torch weaving), laser beams, *etc.*, (see Chapter 3). Monitoring the welding geometry requires more sophisticated sensors, like the laser 3D cameras available for example from Meta Systems – UK and Servo-Robot – Canada (see Chapters 3 and 4). By joint geometry we mean information about the gap, cross section area, mismatch and type of joint (fillet, corner, lap, V-groove, butt, *etc.*). With that information, and with the appropriate software and welding models it is possible to perform several adaptations in real time that go beyond the simple joint finding and tracking tasks (here considered as background tasks): *weaving width*, *welding speed*, *wire feed speed*, *arc voltage*, *arc current*, *etc.*, are some examples of welding parameters that should be adapted if a better quality weld is required (see Chapter 4).

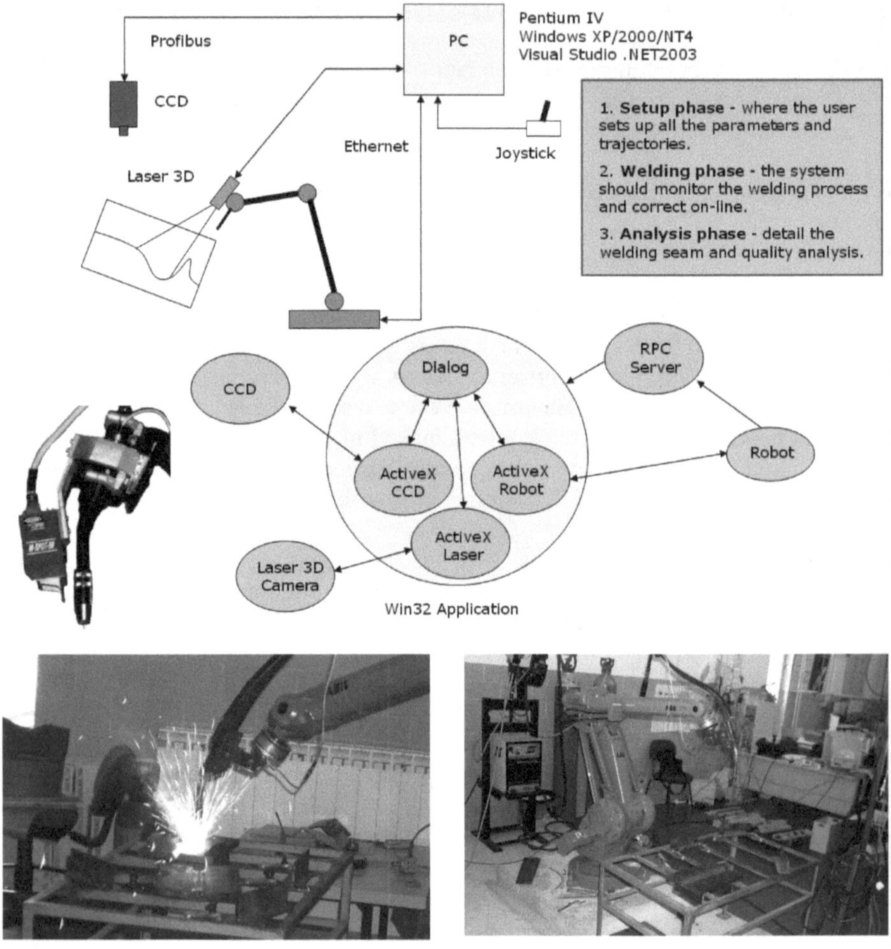

Figure 5.1. Robotic welding system

The software architecture used in this chapter (see Section 4.5) is distributed based on a client-server model that uses software components developed specifically to handle the equipment functionality. Briefly, when we want to use some kind of equipment from a computer we need to write code and define data structures to explore its functionality. We can then pack the software into libraries, which are not very easy to distribute being language dependant, or build a software control using one of the several standard software component technologies available: preferably ActiveX [5] or JAVA [6]. But other technologies could be used; the purpose here is on components and on integration with the environment chosen for operation, not in discussing the possibilities of each technology. Since we use *Microsoft Windows* operating systems, mainly Windows NT/XP and 2000, which are accepted standards in industry, ActiveX is somehow privileged because it was specially built for those environments and is based on DCOM like the operating system.

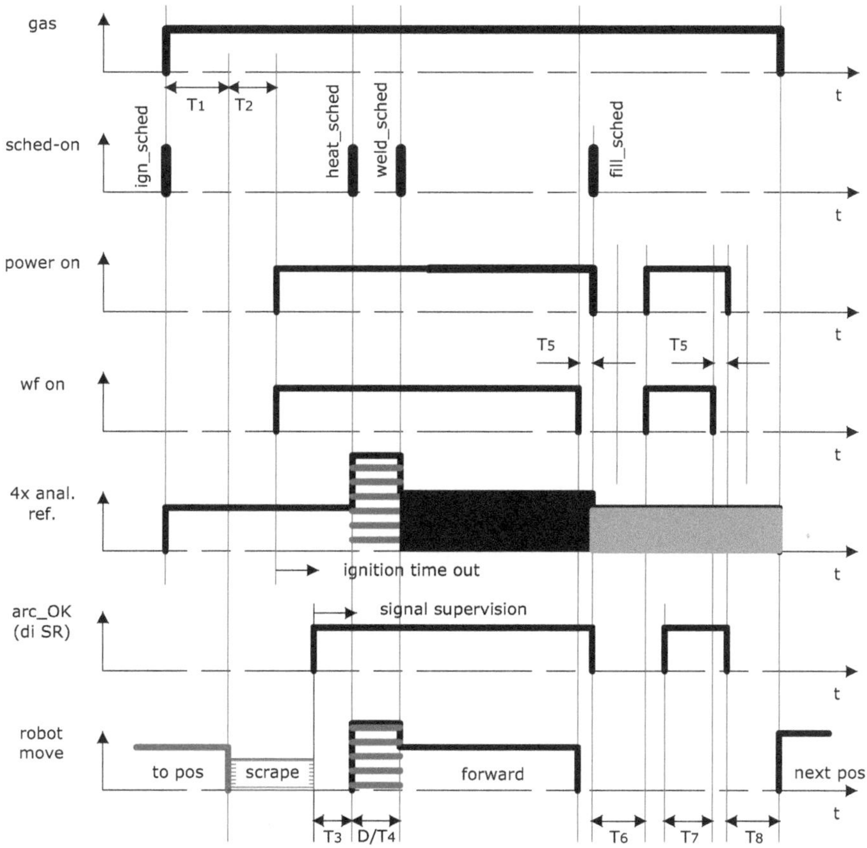

Figure 5.2. Welding sequence implemented by the robot controller (all the timings are programmable by the user)

Using a software control means implementing methods and data structures that hide from the user all the tricky parts about how to have things done with some equipment, focusing only on using its functions in an easy way. Furthermore, those components are easily integrated into new projects built with programming tools that can act as containers of that type of software controls, *i.e.*, they can be added to new projects in a *visual* way. We built several ActiveX software components to use with this project. Those controls reveal to the programmer the basic functionalities of the equipment used: one ABB IRB1400 S4CPlus industrial robot, a SIEMENS VS710 CCD camera and a ServoRobot M-Spot Laser 3D camera. The welding power source (MIG/MAG power source ESAB A350) is controlled from the robot controller using the welding sequence presented in Figure 5.2, and a client-server programming strategy (see Section 4.5.1).

The robot controller software works as a server (Figure 5.3), exposing to the client a collection of services that constitute its basic functionality (see Section 4.5). The robot can start the welding procedure, or terminate it, can be commanded to follow complex trajectories, to simulate the entire process completely or step-by-step, *etc.*, just by answering to remote commands issued from a computer connected to the robot by Ethernet. Basically, the user sends to the robot a complete definition of the welding task including: points, welding parameters (*velocity*, *voltage* and *current*), type of trajectory between and positioning precision, *etc.* All this information is stored in the robot controller and can be used to simulate the welding process and enable any adjustment necessary, or to start/stop the welding process.

```
while never_end;
  switch decision
    case 1: call routine_1; break;
    case 2: call routine_2; break;
    ...
    case n: call routine_n; break;
  end_switch;
end_while;
```

```
PROC weld()
  weldon:=0; i:=1;
  WHILE ((decision1=9401) AND (i<=numberpoints) AND (i>=1)) DO
   weldpoint:=i;
   wd_iref:=trj_voltage{i}; feed_iref:=trj_current{i};
   wd_href:=trj_voltage{i}; feed_href:=trj_current{i};
   wd_ref:=trj_voltage{i}; feed_ref:=trj_current{i};
   IF (trj_type{i}=0) THEN
    weld_on;
    weldon:=1;
   ENDIF
   ppos:=trj{i}; pvel:=trj_vel{i};
   pzone:=trj_prec{i}; ptype:=trj_mode{i};
   move_gen;
   IF (weldon=1) AND ((i+1>numberpoints) OR (trj_type{i+1}=1)) THEN
    weld_off;
    weldon:=0;
   ENDIF
   i:=i+1;
  ENDWHILE
  IF (weldon=1) THEN
   weld_off;
   weldon:=0;
  ENDIF
ENDPROC
```

Figure 5.3. Robot working as a server

5.2.2 CAD Interface

Since the vast majority of companies use CAD software packages to design their products, it would be very interesting if the information from CAD files could be used to generate robot welding programs. That is, the CAD application could be the environment used for specifying the way the welding robots should execute the welding operation on the specified parts.

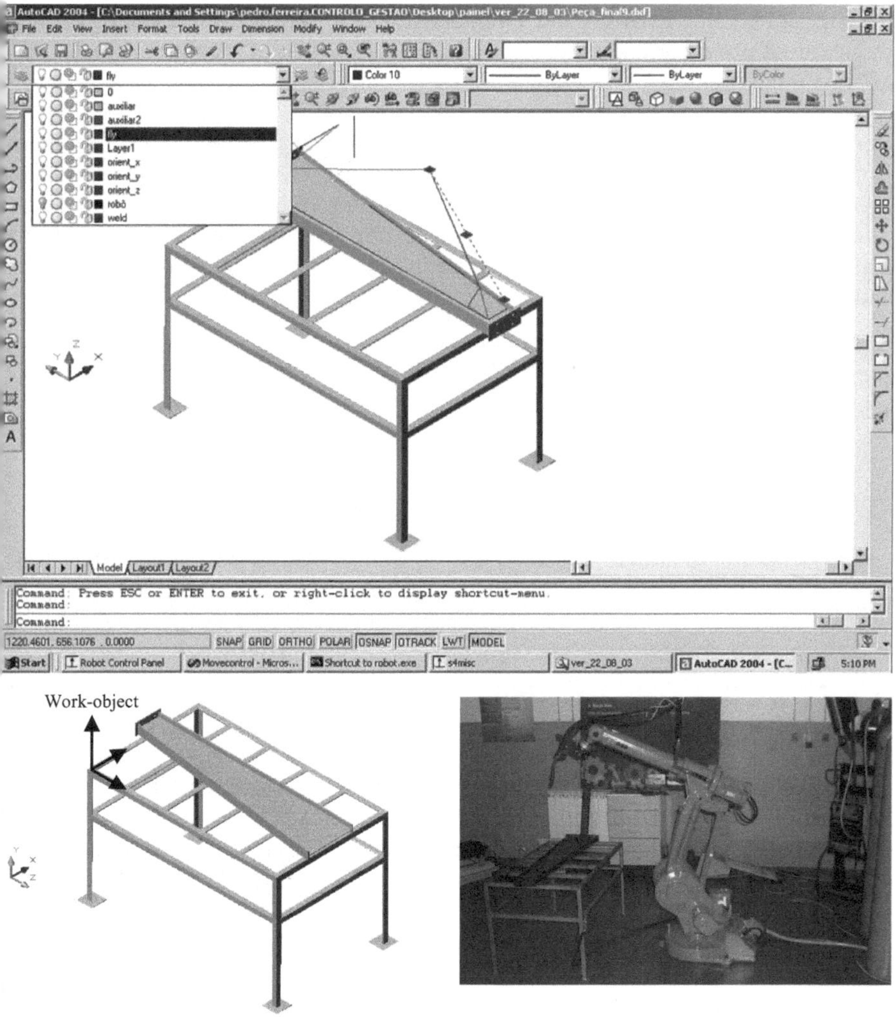

Figure 5.4. Example of welding trajectories using available layers (using AUTOCAD 2004)

Furthermore, since most welding engineers are familiar with CAD packages, this could be a nice way to proceed. The application presented here enables the user to work on the CAD file, defining both the welding path and the approach/escape paths between two consecutive welds, and organize them into the desired welding sequence. When the definition is complete, a small program, written in Visual Basic, extracts motion information from the CAD file and converts it to robot commands that can be immediately tested for detailed tuning. A set of tools is then available to speed up the necessary corrections, which can be made on-line with the robot moving. After a few simulations (with the robot performing all the programmed motions without welding) the program is ready for production. The whole process can be completed in just some minutes to a few hours, depending on the size and complexity of the component to be welded, representing a huge reduction of programming and setup time. Besides, most of the work is really easy off-line programming.

Consequently, having the software tools available and meeting a specific request from industry, a solution was designed using AUTOCAD [7] and DXF files (currently standard files for all CAD software tools). If the user follows some basic rules, and produces a DXF file with all the information needed, then the application developed is capable of extracting that information from the CAD file and is able to produce a robot program almost ready for production.

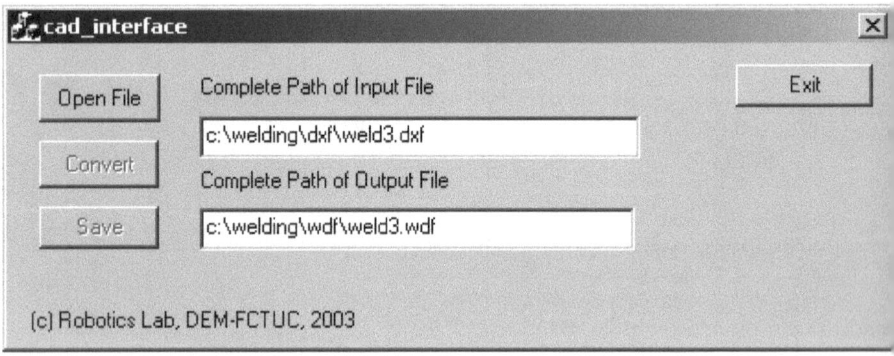

Figure 5.5. Application to extract information from a DXF CAD file

The user starts by having a 3D drawing of the piece to weld and of the table used to hold the piece. The 3D models should be very precise in terms of dimensions and in terms of positioning the welding parts. Then the user should draw and select (Figure 5.4) the sequences and all the trajectories required to weld fully the piece as desired, using the available layers, *i.e.*, using one layer for each trajectory, which is composed by a start-point and an end-point, both with orientation, and the type of motion (welding trajectory or approach/escape trajectory). The welding parameters (*current*, *voltage*, *speed*, *etc.*) are introduced in the selected layer, just by adding labels with the corresponding values. The weld layers should then be renamed for easy identification using a string that starts with the word "WELD".

Simple_Test_Example	Name of the file
4	Number of points in this file
1 - Origem	Name of the point

Definition of point 1

1	Type of point (welding – 0, approach/escape - 1)
656.419922	x (Cartesian position)
-444.451813	y
730.853149	z
0.091980	q1 (quaternion)
0.001690	q2
0.995760	q3
0.002070	q4
0	cf1 (configuration matrix)
-1	cf4
0	cf6
0	cfx
8999999488	ex1 (external axis)
8999999488	ex2
8999999488	ex3
8999999488	ex4
8999999488	ex5
8999999488	ex6
0.00	Current [A]
0.00	Voltage [V]
100	Velocity [mm/s]
5	Precision (mm)
0	Motion type (linear – 0, circular – 1, joints - 2)

...

4 - End

Definition of point 4

1
684.311096
-443.820709
581.514465
0.092050
0.001700
0.995750
0.002130
0
-1
0
0
8999999488
8999999488
8999999488
8999999488
8999999488
8999999488
0.00
0.00
10
5
0

Figure 5.6. Definition of the welding file obtained from the DXF CAD file

The next information is the type of trajectory, to distinguish between weldin trajectories and approach/escape trajectories. After that should be specified th *welding current*, and then the *welding voltage*. Finally, the *welding speed* i specified. All these parameters are separated by spaces, constituting a definitio string. Consequently, a label on a welding trajectory is a string that looks like thi one:

WELD 0 150.0 20.0 10 0

which defines a welding trajectory (0 – welding, 1 – approach/escape, Figure 5.6) with a *welding current* of 150 A, a *welding voltage* of 20 V, a *welding speed* of 1(mm/s and maximum precision in achieving the end-point.

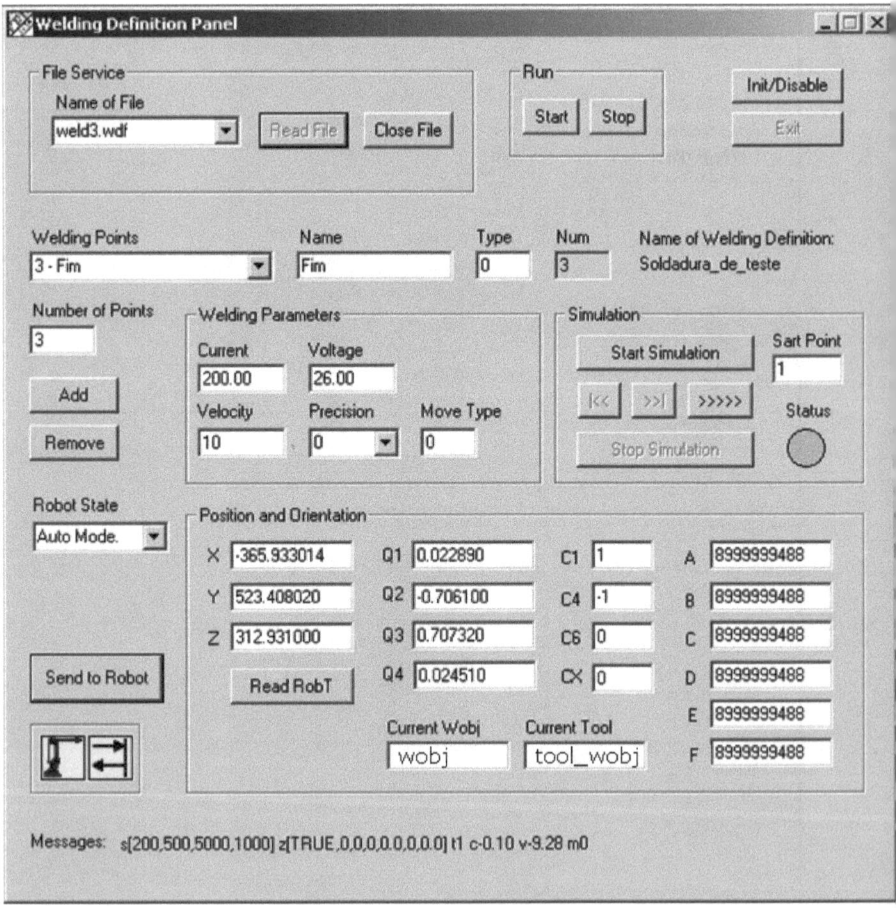

Figure 5.7. Shell of the *WeldPanel* tool

The DXF file generated by the CAD application (AUTOCAD in our case) includes all the information added to specify the welding process. Since the DXF file is an

ASCII file, it is very easy to extract the above-mentioned information, using a simple application (Figure 5.5) that identifies each added welding layer, the trajectories and the related welding parameters, and stores that information in a known way. The definition used here is represented in Figure 5.6.

The generated ".wdf" file is used as input for the application represented in Figure 5.7. This application shows the available definition with the help of several push-down software buttons, and enables the user to change welding parameters, correct points and orientations, simulate the whole process using the real robot and the real piece to weld. The simulation is very realistic, making the final program ready for production. A complete collection of tools was designed to help the user to adjust the points, add extra points, add approach and fly away trajectories, adjust welding parameters, test and simulate the whole process until the operation is as desired. The functions included in these application tools use exclusively the ActiveX control PCROBNET2003 [8], developed by the first author to interface with the RPC services available from the robot controller (see Section 4.5.1).

5.2.3 WeldPanel

With this tool (Figure 5.7) the user can manipulate the welding points that may be obtained initially from a CAD model of the work-piece. Those points may be changed or adjusted, and extra ones may be added as a way to avoid collisions, to optimize trajectories, *etc.*, and to achieve best performance. All points are always referred to the welding torch Tool Center Point (TCP) and to a Work-Object frame defined in the table holding the working object (Figure 5.4). The user may adjust points just by moving the robot to the desired position, a task that can be done from the computer or passing control to the robot *teach pendant*, which is generally easier.

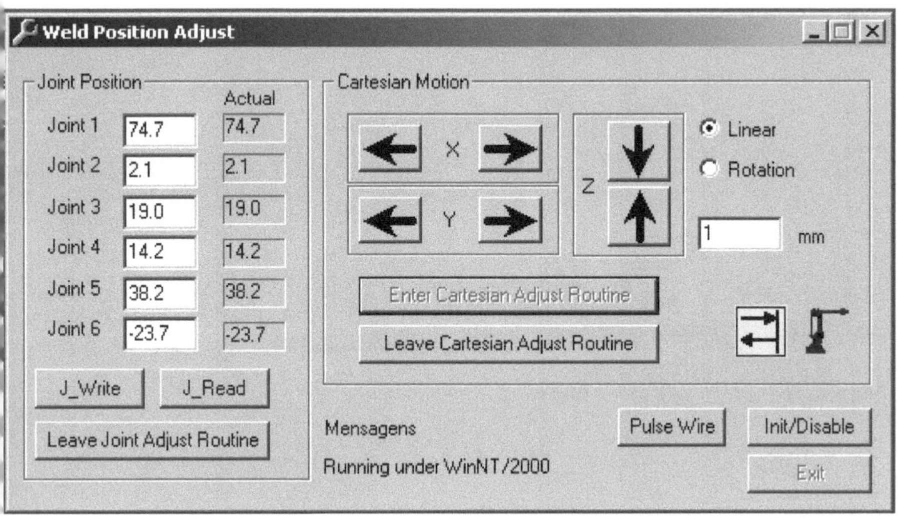

Figure 5.8. Shell of the *WeldAdjust* tool

This "*WeldPanel*" software tool also receives events from the robot, like status changes, actual state of the welding power source and related IO signals, *etc.* The status of the program running on the robot controller and of the network connection is constantly monitored, just to avoid damaging materials and persons but also to prevent system commands on error situations. Those events are simply RPC calls made by the robot controller to an RPC server running on the computer as a system service (see Section 4.5).

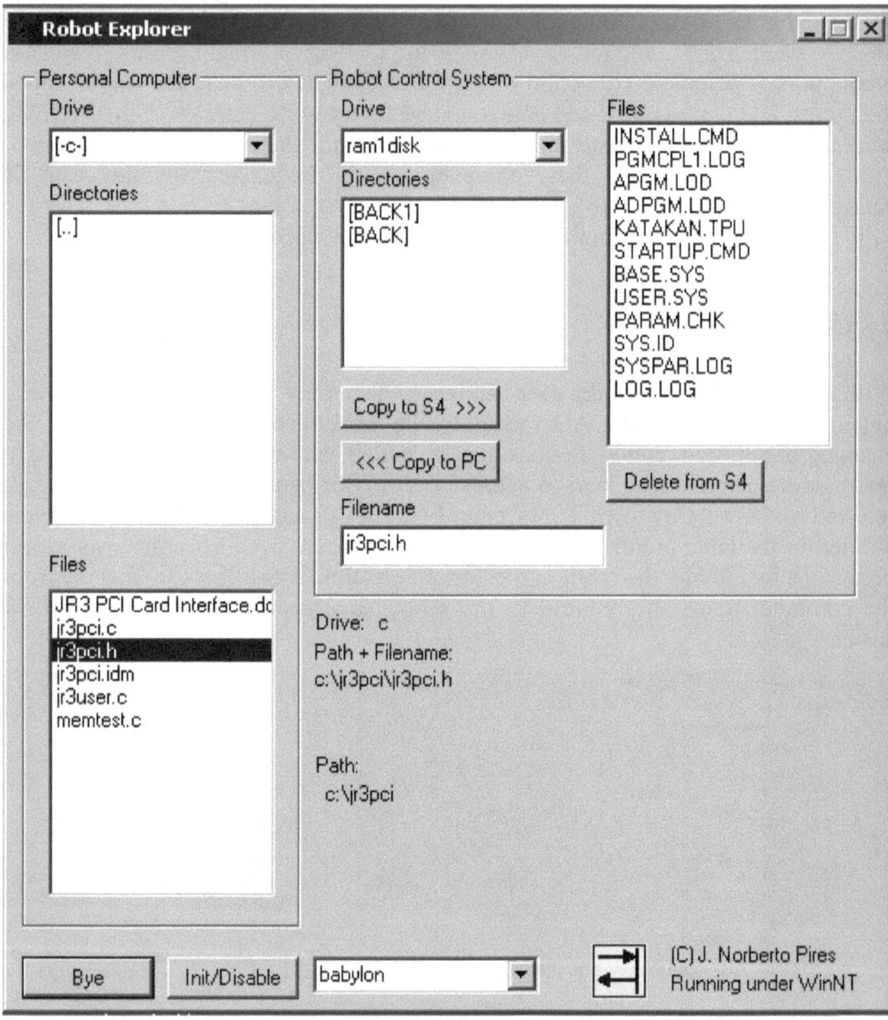

Figure 5.9. Robot File Explorer

5.2.4 WeldAdjust

The "WeldAdjust" software tool (Figure 5.8) is used to adjust points on-line and to acquire points in any given robot configuration and/or any robot program state. Basically it is a robot jogging application that enables the user to position the robot from the computer, using Cartesian XYZ commands or absolute joint commands.

5.2.5 File Explorer

With the "*File Explorer*" (Figure 5.9) software tool the user can exchange files with the robot controller, facilitating the process of transferring programs, modules, *etc.*, to and from the robot controller. It works like the *Microsoft Windows* file explorer, having the available robots as extra "disks". The user can access the robot internal disk and also the external floppy disk, which is implemented using the file access services available from the software component used.

5.2.6 Robot Control Panel and RPC Server to Receive Events

The "*Control Panel*" software tool is used to change the robot controller state and to load and unload modules from the robot controller. The RPC server is used to receive events from the robot controller. As already mentioned events are RPC calls made by the robot controller and fired when pre-programmed actions actually occur. Possible actions include IO change, system state change, program variable change, *etc.* All actions are programmable.

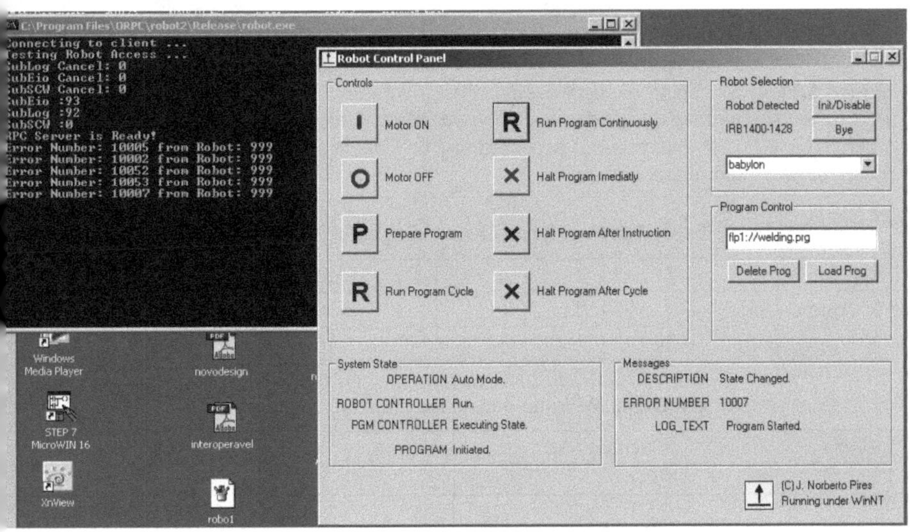

Figure 5.10. Robot Control Panel and RPC server

5.3 Test Cases

Two test cases are presented in this section. The first concerns to the welding o thick plates of a structural steel, where several layers of weld metal need to b deposited in a V-preparation. This situation is very usual in shipyards and companies constructing boilers, pressure vessels or even nuclear components.

Figure 5.11 a, b. Aspect of the working object, welding sequence and obtained weld: **a** - work-piece for multi-pass weld test case (two 20 mm thick plates, 2 mm apart from each other, with a 60º V-groove joint preparation); **b** - layers necessary (welding sequence) to finish the weld and obtained weld

Table 5.1. Position of the welding torch for each layer

Layer	1	2	3	4	5	6	7	8	9
X (mm)	0	0	0	-5	+5	-6	+6	-6	+6
Y (mm)	20	24	28	32	32	36	36	36	36

Table 5.2. Welding data extracted from a database [3]

Layer	Current (A)	Voltage (V)	Speed (mm/s)
1	200	22	5
2	250	28	5
3-9	300	34	5

Distance from the welding torch to the working piece: 17 mm.

The second example approaches the difficulties associated with the welding of fillet welds in complex paths such as those found in the manufacture of metallic beams or trusses.

5.3.1 Test Case 1 – Multi-layer Welding

In this example it is shown how to perform a simple multi-layer weld using the definitions presented in Figure 5.11, and Table 5.1 and Table **5.2**. The number of layers and the placement of each one of them are obtained empirically using charts from *The Welding Institute* [3] and our own experience. The process is performed step-by-step, and any adjustment is introduced in the welding sequence being programmed. Those adjustments can be position adjustments, welding parameter adjustments and introduction or removal of layers. Since the program is stored in a file, it can be used later to weld other similar pieces. The obtained procedure is easy to use and very useful for industrial exploitation because the programmer can easily setup a multi-layer welding procedure controlling and observing the effect of each layer, and acting when necessary. Figure 5.11 shows the working piece, composed by two 20mm thick plates, separated by 2mm, constituting a 60° V-groove joint preparation, and the welding sequence (layers necessary to finish the weld). The position of the torch in each layer is indicated in Table 5.1. The origin of the reference axis system is centered in the bottom of the V-groove. This information is generally not available in the welding databases. The welding data used in this case is indicated in Table 5.2.

5.3.2 Test Case 2 – Multiple Welding Paths

In this example it's shown how to perform a multipoint weld, very common in companies that manufacture metal structures for the construction industry. The idea is to extract points from the CAD model of the piece to be welded. It is usually very simple to build a routine within the CAD software, enabling the user to extract points from the working piece and defining the type of trajectories between those points. This may be the initial procedure, very handy for companies having CAD models of their products. After having the sketch of the definition file, the user must work with it using the *WeldPanel* and *WeldAdjust* tools. The working cycle should result in a properly tuned file for the purpose. An example of the definition file has already been presented in Figure 5.6. The welding parameters may again be obtained from a database. The process can then be simulated for trajectory and welding parameters adjustment, and tested until desirable performance is achieved, including acceptable welding quality.

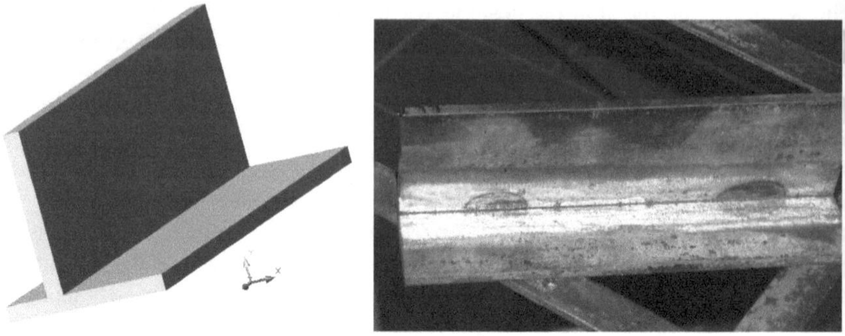

a

b

Figure 5.12 a, b. Aspect of the working pieces: **a** - Fillet weld preparation; **b** - Working table in the laboratory

The on-line remote control of the weld quality is carried out using a vision system focused behind the weld torch, isolated from the torch light. With this method, only surface irregularities of the weld, such as undercuts, underfill or variations in the weld height, may be detected. Other defects like porosity, inclusions or underbead cracks are not detectable using this technique.

The main advantages of the system are the remote control of the welding process and the on-line remote control possibilities of the weld quality.

5.4 Discussion

In this chapter, a robotic welding example was presented, with the objective of demonstrating how the various components of a robotic welding system can be executed. There are several aspects of the presented implementation that are worth discussing in more detail.

5.4.1 IO and Memory Remote Access

To be able to run a robotic welding application with commanding and monitoring functions on a remote computer, the system should be able to:

1. Have IO control capabilities accessible from the remote computer.
2. Have robot memory access capabilities from the remote computer, namely to access program and system variables and data structures.

These features, offered as remote services, are fundamental and must be available. Section 4.5 demonstrates two different examples about how to implement and use those type of services using RPC and TCP/IP sockets. The solution used in this chapter is based on RPC services available from the specific robot controller used (ABB S4CPlus M98). A software component to access those services was designed and used with the examples presented in this chapter, which exposes the complete collection of services available the robot controller just by giving the user the appropriated methods, properties and data structures. For example, admitting that the object is included on a Microsoft Visual C++ .NET 2003 application, the following are possible commands to access IO signals, program variables or system variables.

Write to a digital output named “doGAS”:

```
pcrob.WriteDigital(“doGAS”, value, channel);
```

where “doGAS” is a digital output signal of the robot IO system, *value* is the digital value to write, and channel identifies the RPC socket open with the specified robot controller (the software is able to communicate with any robot on the network through 20 channels for each robot).

Read from the analog input “aiCurrentFeedb”:

```
pcrob.ReadAnalog(“aiCurrentFeedb”, result, channel);
```

where the returned value is stored in the variable *result*.

Write to a numeric variable “decision1”:

```
pcrob.WriteNum(“decision1”, value, channel);
```

where “decision1” is the numeric variable to write and *value* is the desired value.

Read from a position/orientation variable "position":

pcrob.ReadRobTarget("position", result, channel);

where *result* is a data structure of the appropriate type to hold a *robtarge* position/orientation structure [9][10].

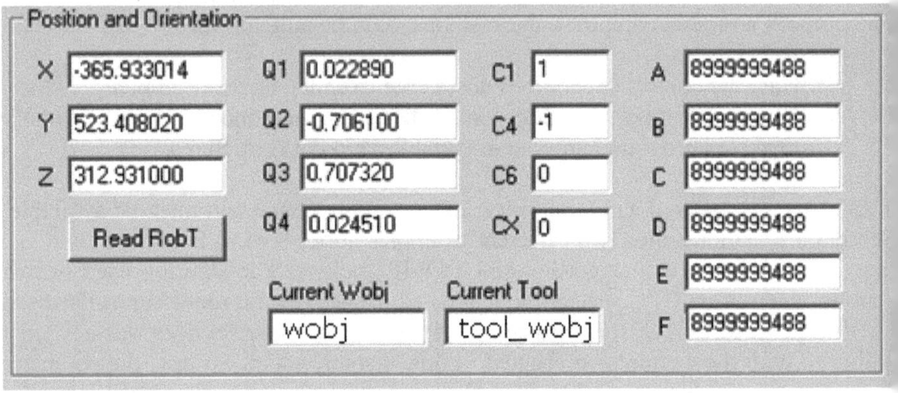

Read actual robot position and orientation

Associated Code

```
m_pon.ReadCurrRobT(&var,&wobj,&tobj);
psa = var.parray;
SafeArrayUnlock(psa);
for (i=0;i<=16;i++)
{
 SafeArrayGetElement(psa,&i,&var1.fltVal);
 warn[i] = var1.fltVal;
}
msg.Format("%.3f",warn[0]); m_x.SetWindowText(msg);
msg.Format("%.3f",warn[1]); m_y.SetWindowText(msg);
msg.Format("%.3f",warn[2]); m_z.SetWindowText(msg);
msg.Format("%.5f",warn[3]); m_q1.SetWindowText(msg);
msg.Format("%.5f",warn[4]); m_q2.SetWindowText(msg);
msg.Format("%.5f",warn[5]); m_q3.SetWindowText(msg);
msg.Format("%.5f",warn[6]); m_q4.SetWindowText(msg);
msg.Format("%.0f",warn[7]); m_cf1.SetWindowText(msg);
msg.Format("%.0f",warn[8]); m_cf4.SetWindowText(msg);
msg.Format("%.0f",warn[9]); m_cf6.SetWindowText(msg);
msg.Format("%.0f",warn[10]); m_cfx.SetWindowText(msg);
msg.Format("%.0f",warn[11]); m_exa.SetWindowText(msg);
msg.Format("%.0f",warn[12]); m_exb.SetWindowText(msg);
msg.Format("%.0f",warn[13]); m_exc.SetWindowText(msg);
msg.Format("%.0f",warn[14]); m_exd.SetWindowText(msg);
msg.Format("%.0f",warn[15]); m_exe.SetWindowText(msg);
msg.Format("%.0f",warn[16]); m_exf.SetWindowText(msg);

SafeArrayDestroy(psa); VariantClear(&var); VariantClear(&var1);
```

Read current position/orientation

Represent obtained data.

Figure 5.13. Code associated with the function Read Actual Position/Orientation

Read actual robot position/orientation:

```
pcrob.ReadCurrRobTarget(result, channel);
```

where *result* is again a data structure of the appropriate type to hold a *robtarget* position/orientation structure [9][10].

The process of accessing information from the robot controller and from the loaded program modules is thus very simple, namely because the necessary data structures are available from the above-mentioned software component. For example, the code associated with the software button "*Read RobT*" of the application "*WeldPanel*" (Figure 5.7) is represented in Figure 5.13.

```
CString msg;
float value = 9301;
nresult = m_pon.WriteNum(LPCTSTR("decision1"),&value);
msg.Format("Error: %d",nresult);
if (nresult != 0) AfxMessageBox(msg);
```

```
CString msg;
float value = 9401;
nresult = m_pon.WriteNum(LPCTSTR("decision1"),&value);
msg.Format("Error: %d",nresult);
if (nresult != 0) AfxMessageBox(msg);
startweld=1;
Sleep(100);
nresult = m_pon.ReadNum(LPCTSTR("weldpoint"),&value);
msg.Format("Error: %d",nresult);
```

```
CString msg;
float value = 123;
nresult = m_pon.WriteNum(LPCTSTR("decision1"),&value);
msg.Format("Error: %d",nresult);
if (nresult != 0) AfxMessageBox(msg);
startweld=0;
```

Figure 5.14. Code associated with some functions of the "*WeldPanel*" application

Figure 5.14 shows the code associated with the software buttons "Start Simulation", "Start" and "Stop" of the software "WeldPanel" tool. Furthermore, Figure 5.15 shows the code associated with the robot jogging capabilities of the software tool "WeldAdjust".

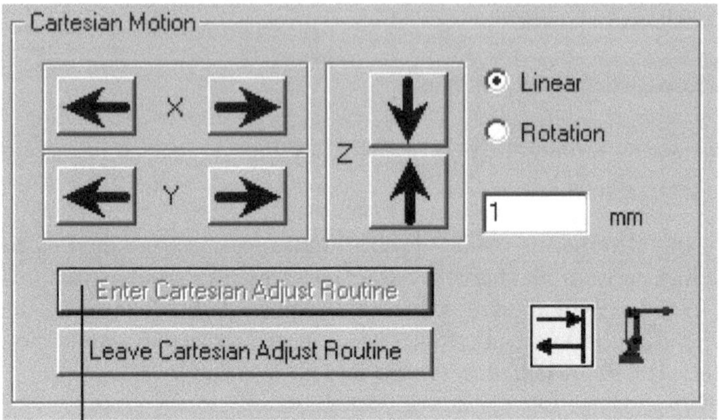


```
float value = 9100;
 nResult = m_pon.WriteNum(LPCTSTR("decision1"),&value);
if (nResult <0) m_msg.SetWindowText("Failled to leave routine.");
if (m_linear.GetCheck())
{
value = 198;
nResult = m_pon.WriteNum(LPCTSTR("varmove"),&value);
if (nResult <0) m_msg.SetWindowText("Failled to leave routine.");
} else
{
 value = 0;
 nResult = m_pon.WriteNum(LPCTSTR("varmove"),&value);
 if (nResult <0) m_msg.SetWindowText("Failled to leave routine.");
}
```

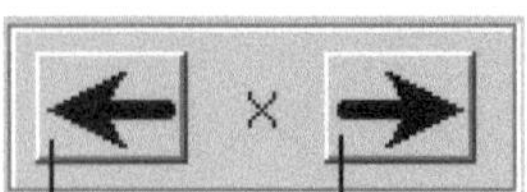


```
CString msg;
float value;
m_inc.GetWindowText(msg);
if (msg.GetLength()) value = -(float) atof(msg); else value = 0;
if ((m_linear.GetCheck() == 1) || (m_joints.GetCheck() == 1)) nResul
= m_pon.WriteNum(LPCTSTR("xx"),&value); else
nResult = m_pon.WriteNum(LPCTSTR("rx"),&value);
if (nResult <0) m_msg.SetWindowText("Failled to leave routine.");
```

```
CString msg;
float value;
m_inc.GetWindowText(msg);
if (msg.GetLength()) value = (float) atof(msg); else value = 0;
if ((m_linear.GetCheck() == 1) || (m_joints.GetCheck() == 1)) nResul
= m_pon.WriteNum(LPCTSTR("xx"),&value); else
nResult = m_pon.WriteNum(LPCTSTR("rx"),&value);
if (nResult <0) m_msg.SetWindowText("Failled to leave routine.");
```

Figure 5.15. Code associated with some functions of the "*WeldAdjustI*" application

.4.2 Software Components

Jsing software components is an interesting solution, because it simplifies the use f certain features hiding the tricky details from the advanced user. Furthermore, ısing "visual" components is also desirable, since they integrate well with the vailable component containers. If the operating system is based on the Distributed COM technology (DCOM), then ActiveX components are one of the obvious olutions.

.4.3 CAD Interface

Actual CAD software packages are powerful 3D tools, very common among nanufacturing companies. These packages are so popular that it's fair to say that lmost every product manufactured in modern countries is designed using some ype of CAD software package. Consequently, using those tools for robot rogramming is desirable, namely on robotic welding applications, since the perator may start the off-line programming of the necessary operations using the D model of the product. The implementation presented in this chapter uses the DXF file standard definition, along with some basic rules, to enable users to add velding information to the CAD file. That information may be automatically extracted from the file and used to program the welding application. This is a traightforward procedure not dependant on the particular CAD software package ısed.

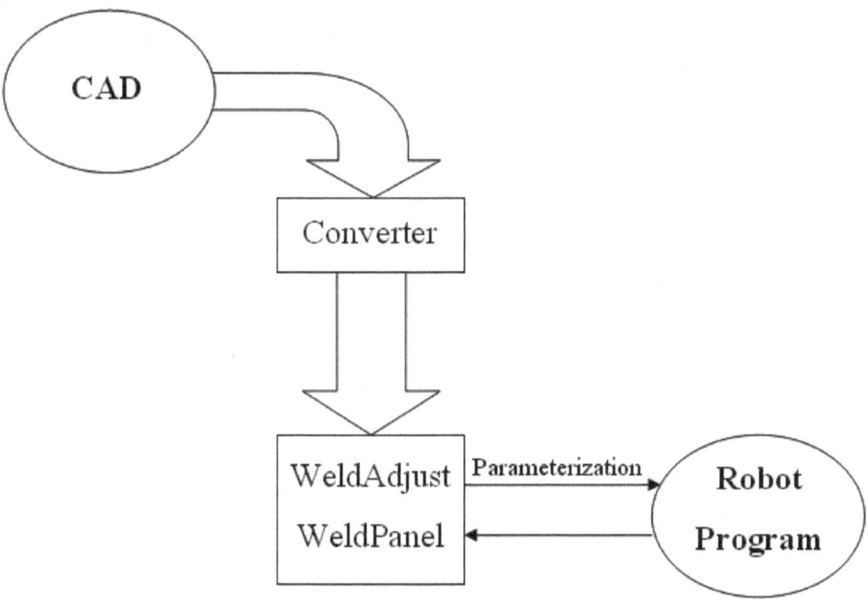

Figure 5.16. Parameterization of an existent welding program

5.3.3.1 Parameterization Approach

With this approach, the welding information, extracted from the CAD model, is used to parameterize a generic already existent robot program, *i.e.*, the welding routines are implemented as general as possible enabling the accommodation of the planned welding tasks (Figure 5.16). In the case presented here the information extracted from the CAD file, and adjusted using the presented software tools, is stored in a ".wdf" file and sent to the robot controller using the option "Send to Robot" of the "WeldPanel" software tool. The information is sent in the form of single column matrices serialized by the sequence that must be followed, *i.e.*, each line of any matrix contains the information correspondent to a certain welding point. As already mentioned, the robot controller is organized as a server, offering a collection of services to the remote computer. Therefore, the following are examples of services implemented in the welding server.

Service 9100 (Move_CRobot): this service is used to move the robot in the Cartesian space with the specified TOOL frame, in accordance with the commanded offsets: x, y, z, rx, ry and rz. Where (x, y, z) are the Cartesian offsets and (rx, ry, rz) are the rotation offsets about the tool axis x, y and z, respectively.

Service 9401 (Welding): this service is used to execute the welding sequence commanded to the robot.

Service 9301 (Simulation): this service is used to execute the welding sequence without igniting the arc, *i.e.*, the welding power source is not activated.

Service 9101 (Move_JRobot): this service is used to move the robot in the joint space in accordance with absolute joint angles commanded from the remote computer.

Consequently, the main routine of the welding server may be implemented as a simple SWITCH-CASE-DO cycle, driven by a variable controlled from the remote computer (Figure 5.17).

Looking into the code in more detail it's easy to find out how it works and how it can be explored, but also how new functionalities can be added into the system. Let's consider for example the *Move_CRobot* service (Figure 5.17) that corresponds to the value 9100 of the variable *decision1*. To move the robot in the Cartesian space the following must be commanded from the remote computer.

1. Enter the service routine: to do that the user must write the value 9100 to the numeric variable *decision1*. The method from the PCROBNET2003 software component used to command that task is

```
pcrob.WriteNum("decision1", 9100, channel);
```

where *channel* identifies the RPC socket open between the robot controller and the remote computer.

```
PROC main()
  TPErase; TPWrite "Welding Server ...";
  reset_signals;
  p1:=CRobT(\Tool:=trj_tool\WObj:=trj_wobj);
  MoveJ p1,v100,fine,trj_tool\WObj:=trj_wobj;
  joints_now:=CJointT();
  decision1:=123; varmove:=0;
  WHILE TRUE DO
    TEST decision1
      CASE 9100:
        x:=0; y:=0; z:=0; rx:=0; ry:=0; rz:=0; move:=0;
        p1:=CRobT(\Tool:=trj_tool);
        WHILE (decision1=9100) DO
          IF (move <> 0) THEN
            p1:=RelTool(p1,x,y,z\Rx:=rx\Ry:=ry\Rz:=rz);
            x:=0; y:=0; z:=0; rx:=0; ry:=0; rz:=0; move:=0;
          ENDIF
          IF varmove <> 198 THEN
            MoveJ p1,v100,fine,trj_tool\WObj:=trj_wobj;
          ELSE
            MoveL p1,v100,fine,trj_tool\WObj:=trj_wobj;
          ENDIF
        ENDWHILE
        decision1:=123; varmove:=0;
      CASE 9101:
        joints_now:=CJointT();
        WHILE decision1=9101 DO
          MoveAbsJ joints_now,v100,fine,trj_tool\WObj:=trj_wobj;
        ENDWHILE
        decision1:=123;
      CASE 9401:
        weld;
        decision1:=123;
        p1:=CRobT(\Tool:=trj_tool);
        MoveJ RelTool(p1,0,0,-200),v100,fine,trj_tool\WObj:=trj_wobj;
      CASE 9301:
        weld_sim;
        decision1:=123;
    ENDTEST
  ENDWHILE
ENDPROC
```

Figure 5.17. Simple welding server running on the robot controller

2. Define the type of motion: the user must specify what type of motion to perform to achieve the target point, *i.e.*, linear motion or coordinated joint motion. This is specified writing to the variable *varmove* (198 for joint coordinated motion and any other value for linear motion). For example, the command

```
pcrob.WriteNum("varmove", 198, channel);
```

specifies joint coordinated motion, using the open RPC socket identified by the parameter *channel.*

3. Command the Cartesian and rotational offsets: the user must write the offsets to the correspondent variables. After that, when the user signals that the offsets are available (writing a value different than zero to the variable *move*), the robot moves to the position/orientation obtained by adding those offsets to the actual position and waits for another motion command. For example, the sequence of commands necessary to move the robot 20 mm in the positive X direction and 10 mm in the negative Z direction should be

```
pcrob.WriteNum("x", 20, channel);
pcrob.WriteNum("y", -10, channel);
pcrob.WriteNum("move", 1, channel);   ←——   robot moves now!
```

where again *channel* identifies the open RPC socket.

4. Leave the service: to leave this service the user must write any value different from 9100 to the variable *decision1*. For example, the following command writes the value -1 to the numeric variable *decision1* and makes the robot program to quit the *Move_CRobot* service:

```
pcrob.WriteNum("decision1", -1, channel);
```

Finally, let's consider the service *Welding* (Figure 5.17) that corresponds to the value 9401 of the variable *decision1*. The simplified version of the code is presented in Figure 5.18.

It is clear from the presented code (Figure 5.18) that the user should command the *Welding* service to execute, after sending the matrices defining the welding sequence. This service commands the robot to follow exactly the command sequence, moving the robot and igniting or stopping the welding arc whenever in the presence of a welding or approach/escape trajectory, respectively.

The presented example shows clearly that there are considerable gains in terms of flexibility and agility when using distributed client-server software architecture to assist industrial welding operations [11]-[15], namely taking advantage of the powerful programming tools developed for personal computers. It also shows that actual CAD packages can be used for robot programming tasks with great advantages, which extend the interest of already largely utilized software tools.

```
PROC weld()
    weldon:=0; i:=1;
    WHILE ((decision1=9401) AND (i<=numberpoints) AND (i>=1)) DO
        weldpoint:=i;
        wd_iref:=trj_voltage{i}; feed_iref:=trj_current{i};
        wd_href:=trj_voltage{i}; feed_href:=trj_current{i};
        wd_ref:=trj_voltage{i}; feed_ref:=trj_current{i};
        IF (trj_type{i}=0) THEN
          weld_on;
          weldon:=1;                                Welding definition
        ENDIF
        ppos:=trj{i}; pvel:=trj_vel{i};
        pzone:=trj_prec{i}; ptype:=trj_mode{i};       Move the robot
        move_gen;
        IF (weldon=1) AND ((i+1>numberpoints) OR (trj_type{i+1}=1)) THEN
          weld_off;
          weldon:=0;
        ENDIF
        i:=i+1;
    ENDWHILE
    IF (weldon=1) THEN
      weld_off;
      weldon:=0;
    ENDIF
ENDPROC

PROC move_gen()
    IF ptype=0 THEN
      MoveL ppos,pvel,pzone,trj_tool\WObj:=trj_wobj;
    ENDIF
    IF ptype=1 THEN
      MoveJ ppos,pvel,pzone,trj_tool\WObj:=trj_wobj;
    ENDIF
    IF ptype=2 THEN
      TPWrite "[MOVE_GEN]: MoveC not implemented.";
    ENDIF
ENDPROC
```

Figure 5.18. Code for the *Welding* service

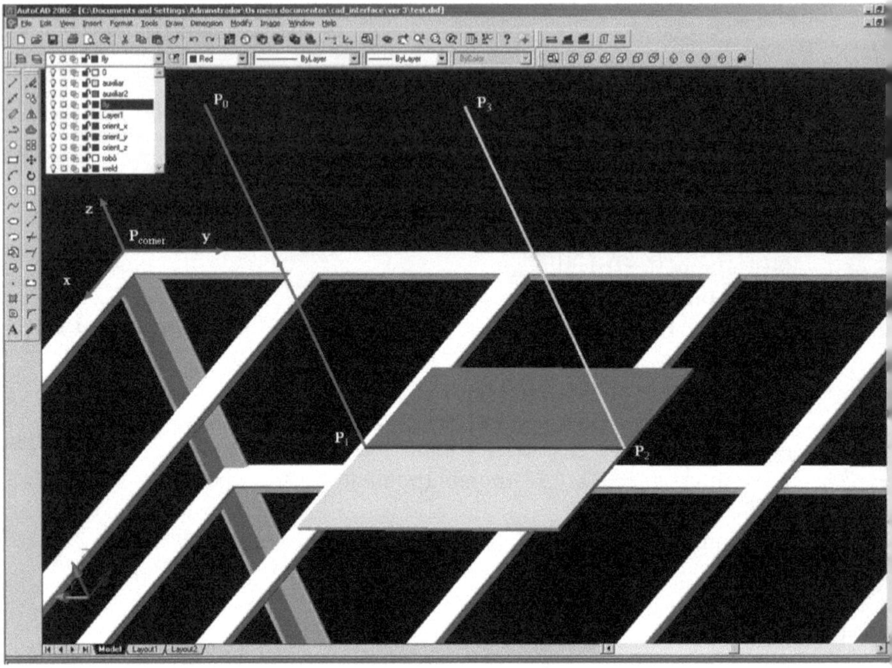

Figure 5.19. Definition of the simple welding example using AUTOCAD

To clarify further, let's consider finally the simple welding example already used in Chapter 4. In that example, the robot is commanded to execute a linear welding on a work-piece placed on a welding table. To demonstrate how this simple task is completely specified and programmed using a CAD package, the welding table and work-piece were modeled in AUTOCAD. The same strategy used before is again utilized to specify points/orientations and trajectories, *i.e.*, they are all defined relative to a work-object point/orientation (or reference system) named P_{corner}. In this way, when exporting points/orientations and trajectories to the robot the only thing needed is a good calibration procedure of the robot TCP relatively to P_{corner} which can be done automatically using sensors (for example, laser position sensors) and special alignment routines, or manually using the robot joystick.

To execute the welding operation it is necessary to specify four points/orientations (P_0 to P_3) and the trajectories between them (Figure 5.19). The following procedures should be used:

1. P_0 should be defined as the approach point/orientation, *i.e.*, a point/orientation that could permit the robot to reach safely the work-piece from the "home" position. P_0 is consequently a non-welding point/orientation and the trajectory to P_0 should be free of obstacles (the user should guarantee that adjusting P_0 accordingly). The precision to reach P_0 should be specified as low.

2. The trajectory from P_0 to P_1 should be defined as an approach linear trajectory, with point P_1 reached with the highest precision at low/medium velocity (let say 100mm/s, for example). Consequently, the label associated with that trajectory (check Section 5.1.2) should be

   WELD 1 0 0 0 100 0

   for an approach/escape trajectory, done at 100mm/s with highest precision in the end-point.

3. The trajectory from P_1 to P_2 should be defined as a welding trajectory with the required welding parameters. For example, the following label could be associated with this trajectory:

   WELD 0 150.0 21.3 10 0

   for a welding trajectory executed at 10mm/s, with highest precision in the end-point, associated with a welding current of 150.0 A and a welding voltage of 21.3 V.

4. The trajectory from P_2 to P_3 should be defined as an approach/escape trajectory done with low/medium velocity without any special precision in the end-point. The following label could be associated with this trajectory:

   WELD 1 0 0 0 100 50

   to specify a trajectory done at 100mm/s, with low precision (50 mm sphere around the selected point).

This information is saved in the CAD file and can be extracted to a ".wtf" definition file, which is used for simulation and final tuning using the already presented tools. Finally, the whole information is sent to the robot using the already presented procedures, based on the routines developed for the robot controller and the "*write variable*" services (see Table 4.1) available from the ActiveX software component used.

5.3.3.2 Code Generation Approach

Another approach would be to generate the welding program, from the scratch, directly from the information extracted from the CAD file containing also the welding specification (Figure 5.20).

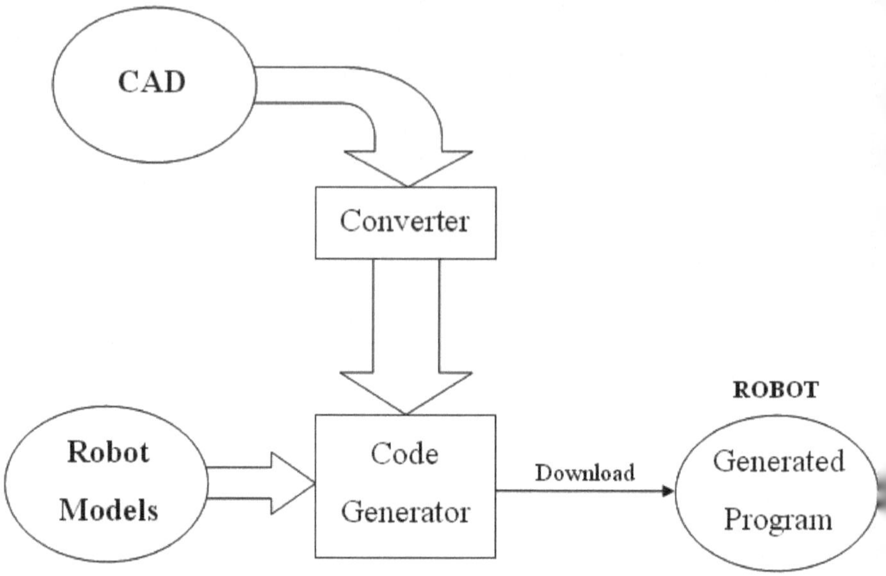

Figure 5.20. Robotic welding: code generation

The first approach is simpler and suits well our demonstrating needs in this chapter. It also works very well for welding applications requiring simple parameterization, or that are constituted by well defined welding trajectories. Nevertheless, for multi-robot solutions and/or applications where it's difficult to typify all the different welding possibilities in terms of trajectories and welding sequences, *etc.*, a code generation based implementation could be desirable. Nevertheless, this possibility requires the existence of code generation modules for each different robot controller used.

5.4.4 Low-level Interfaces for Sensors

The availability of fast low-level interfaces to accommodate tracking sensors is needed, namely laser sensors for seam tracking and on-line welding analysis, which is usually the case for the majority of the industrial robot controllers. Nevertheless, they use proprietary or non-standard protocols which make it hard to attach, program and fully explore those sensors. Consequently, the system low-level interfaces should use well-known data protocols to make the connection to any robot controller standard. For example, if some given sensor is capable of connecting to the robot controller using a TCP/IP socket connection and a proper socket API is available on the robot controller, then connecting any sensor to the robot would be very simple. It would be a question of setting up the client running on the sensor and parameterizing properly the server running on the robot controller (Figure 5.21).

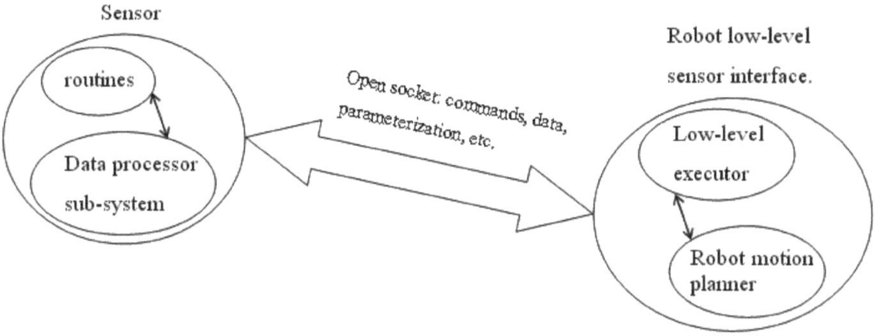

Figure 5.21. Using a TCT/IP socket connection to interface sensors to robot controllers

This concept, extended to offer higher level remote services from the robot controller itself, was presented in Section 4.5.2, where a TCP/IP socket server was designed to work as the prime interface with remote computers for commanding and monitoring applications.

5.5 References

[1] Kusiak A., "Modelling and Design of Flexible Manufacturing Systems", Elsevier Science Publishers, 1986.
[2] Kusiak A., "Computational Intelligence", John Wiley & Sons, 2000.
[3] The Welding Institute, UK, http://www.twi.co.uk
[4] Eddon and Eddon, "Inside DCOM", Microsoft Press, 2000.
[5] Marshall, D., "ActiveX Programming: Building Stable Components using the Microsoft Foundation Classes", Miller Freeman editors, 1998.
[6] Englander, R, "Developing JAVA Beans", O'Reilly Editors, 1997.
[7] KirkPatrick, JM, "The AUTOCAD Book: Drawing, Modeling and Applications Using AUTOCAD 2005", Prentice Hall, 2005.
[8] Pires, JN, "PCROBNET2003, An ActiveX Control for ABB S4 Robots", Internal Report, Robotics and Control Laboratory, Mechanical Engineering Department, University of Coimbra, April 2004.
[9] ABB IRB1400 Users Manual, ABB Flexible Automation, 2005.
[10] ABB RAPID Programming Manual, ABB Flexible Automation, 2005.
[11] Pires, JN, "Interfacing industrial robotic systems with manufacturing tracking and control software. A choice for semi-autonomous manufacturing systems.", Volume 32, nº3, Industrial Robot, An International Journal, MCB University Press., 2005.
[12] Pires, JN, Ferreira, P, Godinho, T, Correia, R," Desenvolvimento de uma aplicação para programação de robôs com interface com ficheiros CAD", Ibero-American Mechanical Engineering Association Journal, Vol. 8, N.3, November 2004.
[13] Pires, JN, "Complete Robotic Inspection Line using PC based Control, Supervision and Parameterization Software", Elsevier and IFAC Journal Robotics and Computer Integrated Manufacturing, Vol.21, Nº1, 2005.
[14] Pires, JN, "Handling production changes on-line: example using a robotic palletizing system for the automobile glass industry", Assembly Automation Journal, MCB

University Press, accepted and will be published in the July 2004 issue (Volume 24 Number 3), 2004.
[15] Pires, JN, and Godinho, Tiago, "Force Control on Industrial Robotics: using a fuzzy P approach", 24, nº1, Sensor Review Journal, An International Journal, MCB University Press, 2004.

What's Next?

In general terms, actual robot controllers have good programming languages although with limited programming tools, which makes the possibility of incorporating remote computer systems very interesting. In fact, this procedure enables the user/programmer to benefit from the advanced programming and analysis tools available for personal computers. This also fits very well with the need to distribute software to the various components of the system, since actual manufacturing tasks (and welding tasks in particular) are generally very complex and require the intervention of several different components: intelligent sensors, robot controllers, computers, software packages, PLCs, *etc.*. This book covered the majority of these aspects, with enough detail, pursuing the basic objective of showing the actual state-of-the-art about welding processes, sensors and systems used to implement robotic welding applications.

Nevertheless, robotic welding remains a very complex task which requires more from the robots and related systems.

Robots need to be cheaper, faster, lighter and much easier to program.

These objectives highlight the standardization of the mechanical and computer platforms used when building robots, as a way to reduce cost and to have robots sharing parts with the other computer systems available on the market.

Furthermore, the above-mentioned objectives also highlight the programming languages, on the human-machine software and hardware interfaces, on the robot controllers, *etc.* And also on the need to better observe and correct in real-time the welding process, which means a better understanding of the welding processes and smarter welding sensors, easier to interface and program.

Moreover, using a robot to execute the welding craft is still a challenge, mainly because the new welding processes (like rapid arc and laser welding) require more efficient systems with higher-levels of adaptability.

Future developments also include the possibility to interface the robot weldin cells directly from advanced 3D CAD packages, which will enable on-lin simulation and testing, also being able to generate complete welding program fully ready for production.

Most of these developments will result from R&D projects done in universitie research institutes and companies, or in cooperation between academia an industry, resulting in technical papers and new products. This subject is perhap one of the most interesting cases of industry-academia cooperation since most o the developments require scientific, technical and operational advances whic require expertise from both worlds.

Robotic welding is therefore an exciting engineering and scientific challenge tha the interested reader can also follow by checking updates, links to leadin institutions and journals, references to selected scientific papers and technica articles, including the software used in this book, along with pictures, videos, *etc.* from

http://robotics.dem.uc.pt/welding_robots

The authors of the "Welding Robots" book will keep this site up-to-date.

J. Norberto Pires
Altino Loureiro
Gunnar Bolmsjö

ndex

Kolibri 360 GmbH
Ferdinand-Jühlke-Straße 7
99095 Erfurt, Deutschland
produktsicherheit@kolibri360.de